IRAN
and the
International Law
of the
Sea and Rivers

by **Bahman Aghai Diba**

PhD International Law

ISBN: 1461009960
ISBN-13: 9781461009962

Introduction

International law of the Sea is among the most important branches of international law and it has made considerable progress in the last decades. This is partly due to the increasing role of the seas and oceans in various aspects of human life and it is also the result of increasing awareness of the states all over the world regarding this branch of international law. The 1982 UN Convention of the Law of the Sea is the most important example of this fact.

This convention is considered a major step away from the older practices of international law as the regulations established by the developed countries of the world and it is a major leap towards general participation of entire world countries to codify an important document of international law. This convention was the result of more than ten years activities of hundreds of lawyers, scientists and diplomats of the world.

In its 50th anniversary of its establishment, the United Nations has called the UN Convention of 1982 on the Law of the Sea as its most important achievement in the UN's entire history.

Also, the country of Iran has long coasts in the Persian Gulf, Sea of Oman and the Caspian Sea. These create special interest and at the same time, special problems for Iran in relation to the law of the sea. At the same time, Iran has important issues related to the international rivers in its borders that have continually impacted Iran's situation in the region and the world.

The current collection is a collection of articles relating to various problems of Iran in the maritime areas and also regarding the international rivers. Some of these articles have been posted in various sites on different occasions.

I take this opportunity to thank Mr. Alireza Ghange Danesh, the respected director of the Ghange Danesh Publications that has published in Iran some of these articles.

Contents

UN 1982 Convention on the Law of the Sea and Iran

The 1982 United Nations Convention on the Law of the Sea (UNCLOS) is considered as one of the most important international legal documents in the world. The Convention which is almost a thousand pages is the result of more than 10 years deliberations of thousands of scientists, policy makers and diplomats and legal experts from a large number of the countries in the world. The convention is a serious effort for participation of as many countries as possible in the regulation of the seas and maritime affairs in the world and it has a special significance for the third world countries because of their new role in codification and formation of the international law.

Iran has been an active participant in the entire sessions of the Third United Nations Conference on the Law of the Sea which culminated in UNCLOS. Iran signed the 1982 Convention in same year, but it has not ratified it. Ratification process in Iran requires that the convention to go through the parliament, but various administrations in Iran since then (1982) have never come up with a plan for the ratification of the convention. There are multiple reasons for the hesitation of Iran. At the same time, noting that Iran has not expressed its inclination to stop joining the parties to the UNCLOS, it is under commitment not to act against its provisions (1).

The best place to understand reasons of Iran for hesitation in joining the UNCLOS is the declaration of the Iranian delegation at the time of signing the 1982 convention in Montego Bay, Jamaica (the Convention is also called the Montego Bay Convention). Although the convention has not welcomed reservations to the provisions of it, but the article 310 of the UNCLOS has opened the door for expressing the concerns of the states who wish to join it. Also, couple of other articles of the convention asks for the expression of the state intentions.

Article 310 of the Convention allows States and entities to make declarations or statements regarding its application at the time of signing, ratifying or acceding to the Convention, which do not purport to exclude or modify the legal effect of the provisions of the Convention.

Article 310 reads:

"Article 310. Declarations and statements "Article 309 does not preclude a State, when signing, ratifying or acceding to this Convention, from making declarations or statements, however phrased or named, with a view, inter alia, to the harmonization of its laws and regulations with the provisions of this Convention, provided that such declarations or statements do not purport to exclude or to modify the legal effect of the provisions of this Convention in their application to that State."

Also, Article 287 of the UNCLOS, paragraph 1, provides that States and entities, when signing, ratifying or acceding to the Convention, or at any time thereafter, may make declarations specifying the forums for the settlement of disputes which they accept.

Article 287, paragraph 1, reads:

"Article 287. Choice of procedure "When signing, ratifying or acceding to this Convention or at any time thereafter, a State shall be free to choose, by means of a written declaration, one or more of the following means for the settlement of disputes concerning the interpretation or application of this Convention:

(a) the International Tribunal for the Law of the Sea established in accordance with Annex VI;
(b) the International Court of Justice;
(c) an arbitral tribunal constituted in accordance with Annex VII;
(d) a special arbitral tribunal constituted in accordance with Annex VIII for one or more of the categories of disputes specified therein."

In addition, article 298, paragraph 1, allows States and entities to declare that they exclude the application of the compulsory binding procedures for the settlement of disputes under the Convention in respect of certain specified categories kinds of disputes. Article 298, paragraph 1, reads:

"Article 298. Optional exceptions to applicability of section 2
"1. When signing, ratifying or acceding to this Convention or at any time thereafter, a State may, without prejudice to the obligations arising under section 1, declare in writing that it does not accept any one or more of the procedures provided for in section 2 with respect to one or more of the following categories of disputes:
(a) (i) disputes concerning the interpretation or application of articles 15, 74 and 83 relating to sea boundary delimitations, or those involving historic bays or titles, provided that a State having made such a declaration shall, when such a dispute arises subsequent to the entry into force of this Convention and where no agreement within a reasonable period of time is reached in negotiations between the parties, at the request of any party to the dispute, accept submission of the matter to conciliation under Annex V, section 2; and provided further that any dispute that necessarily involves

3

the concurrent consideration of any unsettled dispute concerning sovereignty or other rights over continental or insular land territory shall be excluded from such submission;

(ii) after the conciliation commission has presented its report, which shall state the reasons on which it is based, the parties shall negotiate an agreement on the basis of that report; if these negotiations do not result in an agreement, the parties shall, by mutual consent, submit the question to one of the procedures provided for in section 2, unless the parties otherwise agree;

(iii) this subparagraph does not apply to any sea boundary dispute finally settled by an arrangement between the parties, or to any such dispute which is to be settled in accordance with a bilateral or multilateral agreement binding upon those parties;

(b) disputes concerning military activities, including military activities by government vessels and aircraft engaged in non-commercial service, and disputes concerning law enforcement activities in regard to the exercise of sovereign rights or jurisdiction excluded from the jurisdiction of a court or tribunal under article 297, paragraph 2 or 3;

(c) disputes in respect of which the Security Council of the United Nations is exercising the functions assigned to it by the Charter of the United Nations, unless the Security Council decides to remove the matter from its agenda or calls upon the parties to settle it by the means provided for in this Convention." (2)

Noting the above mentioned articles that open the way for the statements in any stage, the Iranian delegation has made a very careful statement at the time of signing the convention. The text of this statement is as follows:

"Iran: Upon signature (10 December 1982):

Interpretative declaration on the subject of straits

"In accordance with article 310 of the Convention on the Law of the Sea, the Government of the Islamic Republic of Iran seizes the opportunity at this solemn moment of signing the Convention, to place on the records its "understanding" in relation to certain provisions of the Convention. The main objective for submitting these declarations is the avoidance of eventual future interpretation of the following articles in a manner incompatible with the original intention and previous positions or in disharmony with national laws and regulations of the Islamic Republic of Iran. It is the understanding of the Islamic Republic of Iran that:

1) Notwithstanding the intended character of the Convention being one of general application and of law making nature, certain of its provisions are merely product of *quid pro quo* which does not necessarily purport to codify the existing customs or established usage (practice) regarded as having an obligatory character. Therefore, it seems natural and in harmony with article 34 of the 1969 Vienna Convention on the Law of Treaties, that only states parties to the Law of the Sea Convention shall be entitled to benefit from the contractual rights created therein.

The above considerations pertain specifically (but not exclusively) to the following:

– The right of Transit passage through straits used for international navigation (Part III, Section 2, article 38).

– The notion of "Exclusive Economic Zone" (Part V). - All matters regarding the International Seabed Area and the Concept of "Common Heritage of mankind" (Part XI).

2) In the light of customary international law, the provisions of article 21, read in association with article 19 (on the Meaning of Innocent Passage) and article 25 (on the Rights of Protection of the Coastal States), recognize (though implicitly) the rights of the Coastal States to take measures to safeguard their security interests including the adoption of laws and regulations regarding, *inter alia* , the requirements of prior authorization for warships willing to exercise the right of innocent passage through the territorial sea.

3) The right referred to in article 125 regarding access to and from the sea and freedom of transit of Land-locked States is one which is derived from mutual agreement of States concerned based on the principle of reciprocity.

4) The provisions of article 70, regarding "Right of States with Special Geographical Characteristics" are without prejudice to the *exclusive right* of the Coastal States of enclosed and semi-enclosed maritime regions (such as the Persian Gulf and the Sea of Oman) with large population predominantly dependent upon relatively poor stocks of living resources of the same regions.

5) Islets situated in enclosed and semi-enclosed seas which potentially can sustain human habitation or economic life of their own, but due to climatic conditions, resource restriction or other limitations, have not yet been put to development, fall within the provisions of paragraph 2 of article 121 concerning "Regime of Islands", and have, therefore, full effect in boundary delimitation of various maritime zones of the interested Coastal States.

Furthermore, with regard to "Compulsory Procedures Entailing Binding Decisions" the Government of the Islamic Republic of Iran, while fully endorsing the Concept of settlement of all international disputes by peaceful means, and recognizing the necessity and desirability of settling, in an atmosphere of mutual understanding and cooperation, issues relating to the interpretation and application of the Convention on the Law of the Sea, at this time will not pronounce on the choice of procedures pursuant to articles 287 and 298 and reserves its positions to be declared in due time." (3)

According to this declaration, the concerns of Iran are focused on several items:

1- The Passage from international straits. The Iranian concern here is on the passage from the Strait of Hormuz in the connecting point of the Persian Gulf to the Sea of Oman. One of the important subjects discussed during the various sessions of the UN Third Conference on the Law of the Sea was the regime of passage from the international straits such the Hormuz strait. The 1982 convention created and approved a new notion for passing from these straits that is called transit passage and it gives more rights

and freedoms to the passing ships than the previous customary regime of passage from these straits. What the Iranian delegation wanted to make clear in the signing of the
convention was that the new rights were based on the contract and therefore they extended only to those who accept
all commitments coming from the 1982 convention and it
did not extend to those who are not the members. The Iranian concern in this case, contrary to the well known idea
that it was against the big naval countries, was also coming
from its conflicts with the Arab neighboring countries that
tried to undermine the rights of the coastal states of such
waterways as much as possible. The Iran-Iraq- war (1980-
1988) had added fuel to this kind of thinking.

During the Third UN Conference on the law of the Sea, the
issue of passage from the international straits had gained special
importance because:

a- It was a matter of controversy between the countries
bordering the straits and other countries especially
the countries with big naval and merchant fleets.

b- The practice of 12 mile territorial sea was recognized and supported by the countries and it added
seriously to the number of international straits that
were less than 24 mile miles and therefore entirely
within the territorial limits of the bordering states.

c- The previous customary regime of passage on the
basis of 1958 convention of the territorial sea (Geneva Convention) was innocent passage. "Under the
regime of innocent passage codified in Section III of
the 1958 Geneva Convention, the rule is established
that transit is innocent only "so long as it is not prejudicial to the peace, good order or security of the
coastal state." The last section of the article also requires that submarines exercising the right of innocent passage navigate on the surface, showing their
flag. In Article 16 a coastal state is given the right to

"take the necessary steps in its territorial sea to prevent passage which is not innocent." This phraseology is nebulous enough as it stands; furthermore, the use of the word "prejudicial" suggests that an actual injury to peace, good order, or security need not be taking place for the passage to be deemed no longer innocent. If a reasonable chance exists that such injury may be in the offing, the coastal state would be in a strong position to decide that the passage is not innocent and exclude the vessel from its territorial waters." (4)

2- The passage of the naval units from the territorial sea or in others words, extending the right of innocent passage to the warships was a controversial issue during the Third UN Conference on the Law of the Sea and after the conclusion of the UNCLOS. The military issues were not in the agenda of the Third UN Conference on the Law of the Sea. However, during the sessions of the conference there were efforts to inclusion of subjects like peaceful use of the oceans but they were not seriously followed. Therefore, the 1958 Convention on the territorial Sea and the 1982 UNCLOS have not clear regulations about the passage of the naval units from the territorial sea. (5) Iran believes that passage of naval ships of other states from the territorial waters is dependent on prior notification and by observance of the innocent passage requirements. Some other countries that have the same policy are Egypt, Oman and Yemen.

3- The UNCLOS 1982 in the article 69 refers to the right of landlocked states to participate in the exploitation of the living resources of the EEZ (Exclusive Economic Zone). This is discussed in a separate article.

4- The problem of Persian as an enclosed or semi-enclosed sea was followed by Iran but it was eventually dropped to many reasons. This is also discussed in a separate article here.

Notes

(1) According to the article 18 of the convention on the law of treaties:

(2)

"Obligation not to defeat the object and purpose of a treaty prior to its entry into force

A State is obliged to refrain from acts which would defeat the object and purpose of a treaty when:

(a) it has signed the treaty or has exchanged instruments constituting the treaty subject to ratification, acceptance or approval, until it shall have made its intention clear not to become a party to the treaty; or

(b) it has expressed its consent to be bound by the treaty, pending the entry into force of the treaty and provided that such entry into force is not unduly delayed. http://www.worldtradelaw.net/misc/viennaconvention.pdf

(3) http://www.un.org/Depts/los/convention_agreements/convention_declarations.htm

(4) http://www.un.org/Depts/los/convention_agreements/convention_declarations.htm

(5) http://www.airpower.maxwell.af.mil/airchronicles/aureview/1974/sep-oct/lawyer.html

(6) RWG De Muralt, the military aspects of the UN Law of the Sea Convention, Netherlands Law Review, 1985, Vol. XXXII.

Iran and the Strait of Hormuz

Recently, due to the revival of the possibility of new sanctions against the Islamic Republic of Iran, the issue of Iran's reactions including closure of the Strait of Hormuz is being discussed. This is a piece on the legal and practical aspects of this issue. The Government of the Islamic Republic of Iran has claimed time and again in the past that if it is put in serious danger it will close the Strait of Hormuz. This claim has several aspects:

Political- the Strait of Hormuz is the channel for export of the 40 percent of the oil production of the entire area. This means, closing the strait is a declaration of war to the exporting countries (Saudi Arabia, Kuwait, Iraq, Qatar, UAE, and Bahrain) and the importing states (including Japan, and the Western states that depend heavily on the oil from the area).

Military- From military point of view, it needs a great effort to close the strait that its narrowest part of it is 34 miles wide. Iran's OPEC governor, Mohammad Ali Khatibi, has said that any threat against Iran will put the flow of oil from Persian Gulf states at risk. "If there is a threat in our region it will not be just Iran's oil exports that will be affected," Khatibi told Reuters in 2007. "It will affect other producers. The fallout will impact other oil exporters like

Iraq, Kuwait, Saudi Arabia, Qatar and the United Arab Emirates. Any problem created in the region by the US or Israel would pose a threat to 40 percent of the world's traded oil," he added. (1)

The editor of Kayhan Newspaper, Hussein Shariatmadari, has said: "if Iran is put under the UN sanctions, the 5+1 states [i.e. the USA, the UK, France, China, Russia and Germany] would the biggest losers from economic and political points of view." He has added: "closing the Strait of Hormuz can be one of the retaliatory reactions [of Iran]….if we are made subject to the sanctions, then naturally we can use some leverages, such as closing the Strait of Hormuz. Closing the Strait of Hormuz will seriously stop the flow of oil to the industrial states and they will face intolerable conditions. Of course, there are numerous other leverages that will be discussed later." (2)

The Strait of Hormuz is the narrow sea passage that connects the Persian Gulf to the Oman Sea. This is the only sea-passage for the export of oil from the Persian Gulf states. The Iranian forces have done several maneuvers aimed at closing the Strait of Hormuz at the time of crisis and the Western forces in the region (in cooperation with the some of the littoral countries or independently) have conducted several maneuvers aimed at deterring such plans.

The threat to close the Strait of Hormuz (3) has been mentioned time and again by many other figures in Iran (4). The officials of the Islamic Republic of Iran have threatened that if the United Nations Organization adopts more serious sanctions against the regime of Iran (due to the nuclear case of Iran which is currently in the Security Council of the UNO), they may resort to the "oil weapon".

Although actions like establishment of an oil bourse (5) that uses euro instead of dollar for its transactions, has been considered as part of the possible ways of using oil as a weapons, it seems that the "oil weapon" is anything that could be used to stop or hinder the actual flow of the much-needed oil to the international markets. This could include a wide range of actions that intentionally stop or seriously reduce the flow of oil from the oil exporting countries to the main consumption centers. (6)

However, the closing of the Hormuz Strait by the regime of Iran has been constantly mentioned inside and outside of Iran as one of the actions that may create global crisis and start a major conflict in the sensitive Persian Gulf region.

It should be noted that any action by Iran to stop the oil flow from the Persian Gulf countries, by blocking the strait Hormuz, attacking the shipping lines, trying to blow up the pipelines or the production and refinery facilities of the other countries in the region (such as Azerbaijan, Kazakhstan, and Turkmenistan in the Caspian Sea, or Saudi Arabia, Kuwait, Iraq, Qatar, the United Arab Emirates, Oman, Bahrain in the Persian Gulf), will be considered as a serious violation of the international laws and regulations for the concerned states. It would be in practice like giving a declaration of war to them. At the same time, it would be a serious challenge to the interests of the major oil importing states, especially the USA, that according to the Carter Doctrine considers the Persian Gulf as an area of vital interests. The US government feels obliged to stop any serious challenge to its vital interest, by all means, including the military actions.

Kenneth R. Timmerman writes: "...The Iranians also plan to lay huge minefields across the Persian Gulf inside the Strait of Hormuz, effectively trapping ships that manage to cross the Strait before they can enter the Gulf, where they can be destroyed by coastal artillery and land-based "Silkworm" missile batteries. Today, Iran has sophisticated EM-53 bottom-tethered mines, which it purchased from China in the 1990s. The EM-53 presents a serious threat to major U.S. surface vessels, since its rocket-propelled charge is capable of hitting the hull of its target at speeds in excess of 70 miles per hour...." (7)

It seems that Iran has actually enough power to block the Strait of Hormuz by sinking several big ships in the main channels of the traffic (although most of the traffic separation lines are in the side of the Strait, which are technically, i.e. according to the international law of Seas, part of Oman's Territorial waters.) Also, Iran may interrupt the shipping in the Strait of Hormuz by mining the international waterway or directly attacking the target vessels. Such actions are enough to stat a full-scale war between the regime of Iran and all other concerned forces.

This is going to be two-sided weapon. The oil prices over a hundred dollar will have serious consequences for the international economy. However, Iran as an oil exporting country will be in a difficult situation if the export of oil is stopped. "Iran's government derives about 50% of its revenues and most of its foreign currency from oil sales..." (8) Iran is not only relying on the oil revenues for its economy, but also it is an importer of the oil products. Iran has not the capacity to produce enough gasoline for the internal consumption and it has to import a major part of its needs from other countries. (In recent years, Oman has been a major source of gasoline imports for Iran). According to Peter Kiernan: "...although Iran is the second largest producer in OPEC, its domestic refining capacity does not meet local demand, and it must import about 170,000 bpd of gasoline, which costs it as much as $4 billion a year..." (9)

Ilan Berman writes "a closer look indicates that the oil weapon, whether in the form of reductions in Iranian output or military moves in the Hormuz Strait, is likely to be a double-edged sword for the Islamic Republic...in their planning, the Bush Administration and its international partners would do well to take doomsday predictions about Iranian energy leverage with a grain of salt. But they should also be thinking carefully about the economic and political costs of inaction. Simply put, Washington must ask itself whether the world would be better off with a temporary spike in energy prices created by a serious Iran strategy, or with a permanent hike in the cost of doing business in a region dominated by an atomic Islamic Republic." (10)

In the past, some suggestions have been made to reduce the importance of the Strait of Hormuz, such as: reducing reliance of the oil exporting countries of the Persian Gulf on the Persian Gulf (especially the Strait of Hormuz) and persuading the exports from the Red Sea (although it makes the supply far from some of the most important centers of demand such as China and India).

Notes:

(1) (http://www.tehrantimes.com/index View. asp?code=173254)

(2) http://news.gooya.com/politics/ archives/2006/11/054527print.php, dated 11/07/2006.

(3) Http://www.dataxinfo.com/toc.htm, this is a website about the Strait of Hormuz.

(4) Ayatollah Ali Khamenei, Iran's Supreme Leader, has given a warning that Iran would disrupt the oil shipments in the Persian Gulf if the USA "makes a wrong move" in the confrontation with Iran over the nuclear program of Iran. (http://www. telegraph.co.uk/core/Content/displayPrintable. jhtml:jsessionid+NMHIWRKS dated 06/25/2006). Also, on 24th of June (2006), the Iranian oil Minister, Kazem Vaziri-Hamaneh, claimed that Iran may use the oil weapon, if this would serve the national interests. (http://asianews.it/viewp.php?1=6550, dated 27 June 2006). Also, Iranian government spokesperson, Gholam-Hossein Elham has said: "Iran would disrupt oil supplies as the last resort if it were punished over its nuclear program". (http://www. zaman.com/include/yazdir.php?bl=hotnews&alt

(5) =&trh=20060626&hn=34302 dated 06/26/2006).

(6) In this field please refer to "Iran Oil Bourse and Petrodollar Wars", Http://www.iranian.ws/iran_news/publish/article_13228.shtml dated 14th of Feb 2006.

(7) http://www.iranian.ws/iran/publish/printer16643.shtml, Bahman Aghai Diba, Iran and the oil Weapon, Persian Journal, July 11, 2006.

(8) http://www.newsmax.com/script/printerfriendly.pl

(9) ?s=pf&page=http://www.newsmax.co dated March 1, 2006

(10) http://www.itp.net/business/features/print.php?id=4530&prodid=&category=arabianbusiness dated 11th of June 2006

(11) The oil weapon and Iran, http://www.atimes.com/atimes/printN.html

(12)

(13) http://article.nationalreview.com/print/?q=MDY5YjFjNDdiNmRjODilMmY5ZTc1NzZiZ dated June 7, 2006

Iranian Straight Baselines
in the Persian Gulf

According to the general rules of the International Law of the Sea, as codified in 1982 Law of the Sea Convention (which is basically a repetition of the 1958 Convention):

"Except where otherwise provided in this Convention[1982 UNCLOS] , the normal baseline for measuring the breadth of the territorial sea is the low-water line along the coast as marked on large-scale charts officially recognized by the coastal State. In the case of islands situated on atolls or of islands having fringing reefs, the baseline for measuring the breadth of the territorial sea is the seaward low-water line of the reef, as shown by the appropriate symbol on charts officially recognized by the coastal State.

1. In localities where the coastline is deeply indented and cut into, or if there is a fringe of islands along the coast in its immediate vicinity, the method of **straight baselines** joining appropriate points may be employed in drawing the baseline from which the breadth of the territorial sea is measured.

2. Where because of the presence of a delta and other natural conditions the coastline is highly unstable, the appropriate points may be selected along the furthest seaward extent of the low-water line and, notwithstanding subsequent regression of the low-water line, the straight baselines shall remain effective until changed by the coastal State in accordance with this Convention.

3. The drawing of straight baselines must not depart to any appreciable extent from the general direction of the coast, and the sea areas lying within the lines must be sufficiently closely linked to the land domain to be subject to the regime of internal waters.

4. Straight baselines shall not be drawn to and from low-tide elevations, unless lighthouses or similar installations which are permanently above sea level have been built on them or except in instances where the drawing of baselines to and from such elevations has received general international recognition.

5. Where the method of straight baselines is applicable under paragraph 1, account may be taken, in determining particular baselines, of economic interests peculiar to the region concerned, the reality and the importance of which are clearly evidenced by long usage.

6. The system of straight baselines may not be applied by a State in such a manner as to cut off the territorial sea of another State from the high seas or an exclusive economic zone. " (1)

Baselines are fundamental to claims to maritime jurisdiction and, frequently, to the delimitation of maritime boundaries. The primary significance of baselines lies in the fact that they provide a starting point for establishing a coastal State's claims to maritime jurisdiction. While often termed "territorial sea baselines", such baselines are fundamental to claims not only to the territorial sea, but all other maritime zones namely the contiguous zone, continental shelf, exclusive economic zone (EEZ). Consequently, the establishment of the location of a coastal State's baselines is a necessary precursor to defining the limits of its zones of maritime jurisdiction, as it is essential to determine the points from which the specified breadth of such zones are measured. Baselines are also important because, just as baselines provide the starting line for the measurement of maritime zones offshore, equally they also represent the outer limit of a State's land territory or internal waters landward of the baseline. Furthermore, base points along these baselines may be crucial to the delimitation of maritime boundaries with neighboring States, especially those based on the construction of equidistance lines. In many cases coastal States have applied the relevant provisions of international law in a liberal or flexible manner and this has arguably resulted in the drawing of excessive baselines. (2)

Iranian baselines in the Persian Gulf (see the text of Iran's law and the report of the US State Department protest to it which is registered as a UN document in the annex) have been source of problems in the Persian Gulf. Perhaps the most obvious one is the incident that happened in 2008 in the Persian Gulf between the Iranian speed boats and the US naval units near the Strait of Hormuz. (3)

The case dated 6 January 2008, involved three United States naval vessels and five Iranian Revolutionary Guard speedboats in the Strait of Hormuz and it was probably at least partially a result of an ongoing Iranian-US dispute over applicable law of the sea. The US claimed that Iranian speedboats, apparently under the command of the Revolutionary Guard, threatened the US naval vessels while they were engaged in what the US claimed was transit passage.

The Strait of Hormuz is bordered by Iran and Oman. Oman is one of 155 state parties to the 1982 UN Convention on the Law of the Sea (Convention). Iran has signaled its support of the Convention by signing it. However, it has not yet ratified or acceded to it, and is thus is not bound by it. The US has not signed it and is not a party to it. The Convention entered into force on 16 November 1994.

It is important to understand the different interpretations of international law regarding Iran's maritime claims in the Strait. The 1982 Convention allows a coastal State to draw straight baselines along an indented coastline and to claim a 12 nautical mile territorial sea from these baselines. Iran claims straight baselines along its coast bordering the Strait and from these baselines a 12 nm territorial sea encompassing the northern third of the Strait and the in-bound designated sea-lane. It also claims three islands just west of the Strait and 12 nm territorial seas around them encompassing much of the navigable waters and both the inbound and outbound designated sea lanes. Although the exact location of the incident has not been published, at some point in their passage into and through the Persian Gulf US warships apparently must pass through Iranian-claimed territorial sea.

Iran's publicly declared position is that countries which are not parties to the 1982 Convention, like the US, cannot avail themselves of the transit passage regime in the Strait. Iran claims that the innocent passage regime applies to vessels of such States moving through its territorial sea and that passage of all warships through Iran's territorial sea requires prior authorization (the latter provision is not consonant with the Convention). Oman recognizes only innocent passage through the Strait for all countries. For both the innocent passage and transit passage regimes, a coastal State has the right to adopt laws and regulations to enhance safety of navigation and regulation of maritime traffic (Articles 21, 42).

The US does not recognize some of Iran's straight baselines along the Strait, arguing that Iran's coast is not indented nor fringed with islands in that area as required by the Convention to use straight baselines (Article 7). That means that the US does not recognize the full extent of Iran's claimed territorial sea in the Strait. Moreover, it may not recognize Iran's claim to the disputed islands west of the Strait and thus Iran's territorial waters claimed there from (4)

Notes

(1) http://en.wikipedia.org/wiki/Baseline_(sea)

(2) http://www.gmat.unsw.edu.au/ablos/ABLOS-08Folder/Session7-Paper1-Bateman.pdf

(3) http://digitalcommons.pace.edu/cgi/viewcontent.cgi?article=1229&context=lawfaculty

(4) http://www.nautilus.org/publications/essays/napsnet/forum/security/08013Valencia.html

Persian Gulf as
an enclosed or semi enclosed sea

The parts of the Persian Gulf that are beyond the limits of the coastal states sovereignty and jurisdiction and traditional was known as a part of the high sea, at the moment, given the new developments of the international law of the sea is considered as " enclosed or semi enclosed sea".

The article 122 of the 1982 UNCLOS has provided a definition of the enclosed or semi enclosed sea that is exactly compatible with the situation of the Persian Gulf. However, this has not created a special legal regime for the Persian Gulf and it has only asked the littoral states of the enclosed or semi enclosed seas, to implement certain general rules of the law of the sea, voluntarily, in a limited or different way.

What was originally pursued by certain states under the idea of the enclosed or semi enclosed seas was not eventually codified in the form of a special legal regime for the Persian Gulf. At the same time, many parts of that idea were provided through other rules such as the exclusive economic zone. This article tries to review the trend of development of the idea of enclosed or semi enclosed seas during the Third UN Conference on the law of the sea (which lasted almost ten years) and then see the approach of various states to the concept of enclosed or semi enclosed seas, including the states around the Persian Gulf, and especially Iran.

The point that all states are free to use the high seas is accepted by the international law of the sea and it has been clearly mentioned in the 1958 Convention on the High Seas. Regarding definition of the high seas, article one of the Convention provides that:

"The term "high seas" means all parts of the sea that are not included in the territorial sea or in the internal waters of a State."

Article 2 of the same convention declares the high seas open to all states and makes it clear that no country can put a legitimate claim for sovereignty over any parts of the high seas and then continues to say:

"Freedom of the high seas is exercised under the conditions laid down by these articles and by the other rules of international law. It comprises, inter alia, both for coastal and non-coastal States:

(1) Freedom of navigation;
(2) Freedom of fishing;
(3) Freedom to lay submarine cables and pipelines;
(4) Freedom to fly over the high seas.

These freedoms, and others which are recognized by the general principles of international law, shall be exercised by all States with reasonable regard to the interests of other States in their exercise of the freedom of the high seas."

However, the scope of the high seas has become increasingly limited by expansion of the new concepts in the international law of the sea. These developments have been mainly the result of technological advancements and more extensive abilities to use the resources in the seas and at the same time, they are due to the security considerations caused by expansion of naval forces. The expansion of coastal states jurisdiction over the adjacent areas, has reduced the scope of the high seas and led to formation of new rules.

Creation of a kind of compromise between the interests of coastal and non-coastal states, in the security, navigation and economic dimensions, led to emergence of general rules for certain regions. At the moment, a considerable portion of the what was known as the high seas previously, has come under various degrees of sovereignty and jurisdiction of the coastal states under such titles as continental shelf, exclusive economic zone (or exclusive fishing zones, EEZ and EFZ respectively), archipelagic waters, adjacent area and so on.

In the same context, the coastal states of enclosed or semi enclosed seas have made a claim the special circumstances of such seas require determination of special methods for implementation of the general rules of the international law of the sea. (1) But this did not lead to emergence of a new and coherent legal regime for the enclosed or semi enclosed seas.

The above said claims were initially made and followed during the numerous sessions of the Third United Nations Conference on the Law of The Sea (1973-1982) which culminated in conclusion of the 1982 UN Convention (2). However, the roots of this idea were already apparent in the first Law of the Sea Conference in 1958 (Geneva). The representative of Iran in the Geneva Conference (3) has said: "the Geneva Conventions are arranged more to the favor of the states surrounding oceans than the states around enclosed or semi enclosed seas." (4)

Several other states have made points regarding the enclosed or semi enclosed seas and argued that the legal status of such regions should be different from the high seas. In the same conference, the delegations of Romania and Ukraine have made remarks about the necessity to establish a special regime for shipping between the high seas and the enclosed seas.

These remarks have been supported by the USSR and at the same time, the USA, and the UK opposed them and eventual nothing in this line was mentioned in the 1958 convention. (5)

the Russians were not happy that the states in the other regions of the world establish a special regime for the enclosed or semi enclosed seas and they were concerned only about the regions under their control. The main reason for such a policy was that the Russian naval fleet was expanding. Even during the Third UN Conference on the law of the sea, the Russians did not follow this idea seriously. (6) The aim of Russians in pursuing this idea was mainly blocking the traffic of other major naval states in the Baltic Sea because the traffic of naval units of the other countries to the Black Sea was already regulated according the Montero Treaty and the Russians were only worried about the situation in the Baltic Sea. However, expansion on the Russian naval power made the travelling of Russian naval units to other places more important than this concern.

During the preparatory activities that were made for holding the Third UN Conference on the Law of the Sea, the Committee on Peaceful Uses of the Seabed and Oceans Floor, on 16 August 1972, discussed the idea of codification of a special set regulations for the enclosed or semi enclosed seas and requested to put it in the agenda of the 3rd UNCLOS. (7)

Now, before we go further in reviewing the stages for codification of this idea in the 3rd UNCLOS, it is necessary to look more carefully in the expression of the enclosed or semi enclosed seas. Some of the experts and writers in the field of the international law, in the past have used the enclosed seas to make a reference to the bodies of water that have no outlet to the high seas, such as the Caspian Sea, Bahrol-mayyaet (the Dead Sea), and Aral. But it is better to call such bodies of water as "lake" and what we have here in mind is the enclosed or semi enclosed seas that we will review its definitions shortly. It is interesting that these two (the enclosed or semi enclosed seas) come together and numerous efforts to find a way to distinguish and separate them have failed. The closest concept to what became known as the enclosed or semi enclosed seas is the concept of "gulf". Therefore, we need to explore a little in the materials related to gulfs in the international law. According to the rules of gulfs and bays in the international law (article 7 of the 1958 Convention on the Territorial Sea and also 1982 UNCLOS) sometimes, a body of water becomes part of the internal waters. This is the case when the entire coasts of a sea belong to a single state and its strait is less than 24 miles. This is called a closed sea. The best example for it is the Azov Sea. Some other states also claim that certain gulfs that do not belong to them according to the general rules of the international law of the sea, but due to historical reasons, they are part of the internal waters. They call these "historical bays."

1958 and 1982 conventions of the law of the sea have no provisions for the historical bays. However, during the 1958 conference some proposals in this field were discussed. The same conference asked the secretariat of the United Nations to conduct a study regarding the regime of historical waters including historical bays. This study was made ready by the UN Secretariat and it was published in 1962. (8) However, no legislation or codification of laws and regulations in this field was made in the international level.

According to the above mentioned study, the customary law may accept the claim of a state regarding historical ownership of a gulf if sufficient evidence can be provided about the effective historical ownership that is not contested by other states. In such cases, the concerned state can draw a line at the entrance of a historical bay and consider it as the baseline for measuring the maritime territories and make the water behind such a line internal water. There are several claims by states regarding historical bays including Russia's claim for Azov Gulf, Canada's claim on Hudson Gulf and the US claim over the Chesapeake Bay (9). All of these are bigger than the gulfs mentioned in the article 7 of the 1958 Convention and article 10 of the 1982 Convention.

What is closer than the historical bays to the concept of the enclosed or semi enclosed seas is the special situation of a body of water that several states surround it and it is connected to the other seas through a strait. The Russian Proavo believes the issues of such regions should be administered by common agreement and while shipping must be free for all, there must be some limitations for the traffic of military vessels and flights over these regions and the resources beyond the territorial sea should be fairly divided. He has mentioned the Back Sea and Baltic Sea as examples of the enclosed or semi enclosed seas. (10) Of course this concept is not generally accepted by many states in the world. (11) It is timely here to refer to the contents of the International Convention for Prevention of oil Pollution from Ships (London Convention of 1973). In this convention, certain environmentally sensitive regions that are threatened by serious pollution are declared as special regions. In the special regions, we can see the characteristics of the enclosed or semi enclosed seas and the Persian Gulf is one of them.

It is also timely to refer to the claims of some states for having exclusive ownership over a section or entire area of seas and even oceans, no matter how many other states have surrounded those bodies of water. These are historical claim that have not affected the new developments in the international law of the sea. Among such claims we can refer to the claim of Genoa over the Ligorian Sea, claim of Venice over the Adriatic Sea, Norwegian claim over the northern section of the North Sea, claims of Scotland and later the Britain over their adjacent seas, and claims of Spain and Portugal over sectors of oceans. These claims are not supported by the important writers of the international law and as mentioned, in the latest developments of the international law of the sea, they have been ignored totally. (12)

Talking about the process of development of the idea of the enclosed or semi enclosed seas, before the 3rd UNCLOS, it is necessary to mention the efforts of Professor Alexander, in the US Rhode Island Law of the Seas Institute, who has made clear points about) the enclosed or semi enclosed seas in 1973. (13) He has mentioned in his works that the characteristics of) the "semi-enclosed seas" are as follows:

a) The area of it must be at least 50 thousands square miles

b) It must be a part of an ocean and not a part of another semi-enclosed sea

c) A strait connects it to the high seas and this strait should not be more than 20 per cent of the entire area of the concerned body of water

He also had proposed to the United Nations to take note of:

A) Definition of the enclosed or semi-enclosed seas

B) Regulations of various regions for security of shipping

C) Cooperation of the coastal states of the semi-enclosed seas for shipping

D) Cooperation of the littoral states for combat against pollution with the assistance on international organizations

E) Assistance of the United Nations to the littoral states of the semi-enclosed seas for settlement of disputes on delimitation of maritime territories.

The enclosed or semi-enclosed seas in the 3rd UNCLOS

During the 3rd UNCLOS, especially its second session (Caracas 20 June to 29 August 1974), a considerable number of the representative delegations took part in the discussions on the enclosed or semi-enclosed seas. In the second committee of the session, a set of draft articles were presented in this field for inclusion in the final convention. (14) the differences of views between the littoral states of the enclosed or semi-enclosed seas, with each other and with other states who were concerned about their interests in such regions was really extensive. The conflicts between the enclosed or semi-enclosed seas became more visible during the next sessions of the conference and they took different positions regarding the issue. The littoral states of the enclosed or semi-enclosed seas due to various reasons were seeking a special regime for such regions. Some of them were only after a special arrangement for shipping (like Iran), some others were after the idea due to seeking a special arrangement for delimitation of marine areas (such as Iraq), and yet others were looking for a special regime for fishing rights. Eventually, it was make clear that there is no common point between the demands of these states and for the same reason no special regime for the enclosed or semi-enclosed seas emerged.

During the Caracas session (1974), representatives of Iran presented various suggestions regarding the enclosed or semi-enclosed seas. The Iranian delegation, inter alia, said:

"The special geographical circumstances of the enclosed or semi-enclosed seas require that the general rules that are globally used, are not applied in these cases. It is better to have regional and bilateral arrangements in such sea." (15)

Also, they said:

"The Convention (of the law of the sea) should give more competence to the littoral states of the enclosed or semi-enclosed seas for preventive and limiting measures under regional arrangements for such seas... one of these areas is protection of the natural resources of these regions that are vulnerable, and the other point is maritime research. This kind of researches should be done only after getting the special permission of the coastal states of the enclosed or semi-enclosed seas. "(16)

During the discussions on the marine pollution, the representative of Iran expressed hopes that "a fund is established for compensation of pollution damages." (17)

At the session, the representative of Algeria believed that certain general rules should be established for the enclosed or semi-enclosed seas, but the details should left to the littoral states of such regions to determine. Yugoslavia supported inclusion of specific rules for the enclosed or semi-enclosed seas in the final convention. The representative of this country in the fourth session of the 3rd UNCLOS said: " it is necessary to mention specific rules due to main reasons: first, complexity of these regions for shipping and navigation, and their connection to other seas through a strait and second, various kinds of pollutions and the low speed of water circulation in these regions and the necessity to take special measures for their protection and management of the living resources due to the geographical characteristics." (18)

In the Geneva meeting of the 3rd UNCLOS (17 March to 9 May 1975), the issue of the enclosed or semi-enclosed seas was discussed and an informal consultative group was formed to follow it. (19) Eventually in the draft text of the convention which was prepared by the fifth session of the conference in New York (12 August to I September 1976), and it was called "Informal Single Negotiating Text", and was the basis for all later sessions of the conference, there was a section for the enclosed or semi-enclosed seas. The chapter 10 of this text contained articles 133 to 135 about the enclosed or semi-enclosed seas. The text had a definition of the enclosed or semi-enclosed seas and a section about cooperation of the littoral states of the enclosed or semi-enclosed seas.

After this stage, in the later sessions, although the issue of the enclosed or semi-enclosed seas was there, but the discussions mainly took place in the informal meetings. Following the fifth session of the conference, the second committee (which covered the discussions related to the enclosed or semi-enclosed seas), decided to follow only certain subjects in the public sessions and the issue of the enclosed or semi-enclosed seas, was not one of them. Therefore, the portion of the "Informal Single Negotiating Text" which was related to the enclosed or semi-enclosed seas did not see much change until the final stages of the conference.

What had been mentioned in the "Informal Single Negotiating Text", did not provide an exact definition of the enclosed or semi-enclosed seas and at the same time, issues of interest for several states (including and especially Iran), meaning the issue of shipping in the enclosed or semi-enclosed seas, flight over them, and problems related to the strait that connects these regions to the others, was not there at all. Article 133 of the "Informal Single Negotiating Text" provided:

"In this chapter, the expression of the enclosed or semi-enclosed seas means gulfs, and regions or seas that two or more states are surrounding it, and it is connected through a narrow outlet to the high seas, or entirely or mainly it is the territorial sea and exclusive economic zone of two or more littoral states."

This definition was vague. There were not enough materials in it to distinguish the enclosed and semi-enclosed seas, conditions of the enclosed or semi-enclosed seas and the characteristics of the outlet that connects it to the high seas. Article 134 of the "Informal Single Negotiating Text", had tried to strike a balance between the interests of the littoral states of the enclosed and semi-enclosed seas and others regarding exploitation of the living resources, protection of the marine environment and maritime research. The test provided that:

"The states around the enclosed and semi-enclosed seas will cooperate with each other for implementation of the rights and duties provided in this convention. For this reason, they will act to the following points directly or through the proper regional arrangements:

a- Coordination of management, protection, exploitation and exploration of the living resources

b- Coordination in implementation of the rights and duties related to the protection of the marine environment

c- Inviting properly the other interested states or international organizations for cooperation in advancement of the provisions of this article."

Although several amendments were proposed for this article in various meetings of the conference, only the amendment for the introductory part of this article was welcomed. According to this amendment that was jointly sponsored by Mexico, Tunisia, and Turkey in the introductory section of this article it was said: "the concerned states will try to coordinate the following cases." (20) This amendment was important due to the further weakening the commitments of the littoral states of the enclosed and semi-enclosed seas for cooperation (we will see this more carefully here).

Another article appearing in the "Informal Single Negotiating Text" after the sixth session of the conference on the enclosed and semi-enclosed seas was article 135 which provided:

"The regulations of this section have no effect on the rights and duties of the littoral states or other states according to the present convention and they will be implemented in a way suitable with them."

The problem of definition for the enclosed and semi-enclosed seas

During the 3rd UNCLOS, many states tried to introduce the seas near to them as the enclosed and semi-enclosed seas. The representatives of Iran in one of the sessions said that there were 40 to 50 enclosed and semi-enclosed seas. (21) the bodies of water that were more frequently called as the enclosed and semi-enclosed seas were: Baltic Sea, Persian Gulf, Red Sea, South China Sea, Andaman Sea, and even Mediterranean Sea (and sections of it such as Aegean Sea, Adriatic, and the Black sea).

Some of the states believed that the definition of the enclosed and semi-enclosed seas should not be very strict. The best example was the case of the Mediterranean Sea. The Russians opposed the idea of this case and they argued that the Mediterranean Sea was too large and many states were around it and the sea had been used freely by all stats since long ago. The Russians also insisted that the seas with multiple outlets to the high seas and used for international shipping should not be included in a special regime. (22)

These differences of views reveal the importance of the definition. The representatives of Iran, in order to distinguish between the enclosed and semi-enclosed seas, proposed certain rules that could not get enough support. Those suggestions included:

"a - the expression of enclosed seas is related to small seas that are surrounded by two or several states and are connected to the open seas by a narrow outlet.

b- The expression of semi-enclosed seas applies to a maritime zone that is situated in the periphery of an ocean and surrounded by two or several states." (23)

This proposal contained several vague geographical and legal terms, including "the narrow outlet", periphery of an ocean, open seas, and maritime zone.

Other states did not try seriously to separate these two (the enclosed and semi-enclosed seas). Although the definition mentioned in the Single Negotiating Text was not welcomed by many states, nothing better was found to replace it. Some of the Eastern European states, due to their own conditions, proposed to seas which contain several other enclosed and semi-enclosed seas, should not be considered as the enclosed and semi-enclosed seas. Iraq and several Arab states supported this idea during the fifth and sixth sessions of the 3rd UNCLOS. Turkey, Yugoslavia, Algeria, and Libya proposed in the sixth session to delete the term of "enclosed seas" because its meaning is not clear and keep only the semi-enclosed seas. (24)

However, this was not approved too and eventually what was mentioned in the 1982 UN Convention on the Law of the Sea included both expressions.

Article 122 of the part IX, in the1982 Convention provided the following definition for the enclosed or semi-enclosed seas:

"for the purpose of this Convention, "enclosed or semi-enclosed sea "means a gulf, basin or sea surrounded by two or more States and connected to another sea or ocean by a narrow outlet or consisting entirely or primarily of the territorial seas and exclusive economic zones of two or more coastal states."

Cooperation of the states surrounding the enclosed or semi enclosed seas

Many of the participants in the 3rd UNCLOS, the characteristics of the enclosed or semi enclosed seas made it necessary for the states surrounding such seas to cooperate in prevention of pollution, protection of the living resources and creation of safety for shipping. For the same reason, in the draft of the single negotiating text, the issue of cooperate between the coastal states of the enclosed or semi enclosed seas in prevention of pollution, protection of the living resources and creation of safety for shipping and also scientific research were mentioned and no state opposed to it.

There were some suggestions for amendment of the articles in the concerned chapter for the enclosed or semi enclosed seas, but none of them gained sufficient endorsement. During the fourth session of the 3rd UNCLOS, the UAE proposed that the objective of cooperation between the coastal states of the enclosed or semi enclosed seas must be focused on preventing the negative effects of over fishing and also supporting the fishers' communities. Also, Finland proposed that the coastal states of the enclosed or semi enclosed seas should be made committed to conclude regional agreements for protection of the living sources so that the difficulties resulting from creation of the EEZ and EFZ for the traditional activities of the fishermen is removed.

Iraq proposed to the fourth session that the cooperation of coastal states of the enclosed or semi enclosed seas should result in equitable sharing in the living resources. Iraq amended its stance during the sixth session of the 3rd UNCLOS and made its proposal more focused. According to the new and revised Iraqi proposal, the states around the enclosed or semi enclosed seas were obliged to enter into regional agreements for finding equitable ways for the issues related fishing and in the process of this activity, the special circumstances of the states around the enclosed or semi enclosed seas and their population's needs should be taken into consideration. (25)

However, as it was mentioned earlier, these suggestions and similar ones did not get a chance to become part and parcel of the convention and only the common suggestion of Algeria, Libya, Yugoslavia, Romania and Turkey which was based on the previous proposal of Finland was accepted generally. The proposal had mentioned that conclusion of agreements between the concerned states can be one of the ways of realization of cooperation between the littoral states of the enclosed or semi enclosed seas.

Iranian delegation had originally supported the idea of necessity of conclusion of agreements. (26)

Eventually, what was actually mentioned in the 1982 Convention regarding the cooperation of the littoral states of the enclosed or semi enclosed seas was as follows:

"*123- Cooperation of States bordering enclosed or semi-enclosed seas*

States bordering an enclosed or semi-enclosed sea should cooperate with each other in the exercise of their rights and in the performance of their duties under this Convention. To this end they shall endeavor, directly or through an appropriate regional organization:

(a) to coordinate the management, conservation, exploration and exploitation of the living resources of the sea;

(b) to coordinate the implementation of their rights and duties with respect to the protection and preservation of the marine environment;

(c) to coordinate their scientific research policies and undertake where appropriate joint programs of scientific research in the area;

(d) to invite, as appropriate, other interested States or international organizations to cooperate with them in furtherance of the provisions of this article." (27)

Shipping issue

There were many disagreements among the littoral states of the enclosed or semi enclosed seas and also between these states and the non-littoral states regarding navigation in the enclosed or semi enclosed seas. In the Caracas Session of the 3rd UNCLOS, a number of the littoral states of the enclosed or semi enclosed seas proposed that issues relating to the access to such regions should be accompanied by the issues related to the straits that connect such regions to other seas. At the juncture, the width of territorial sea was generally getting expanded. 12 miles had become a regular practice and the adjacent area was claiming another 12 miles for certain jurisdictions of the littoral states and more important than that the idea of 200 miles Exclusive Economic Zone (EEZ) was going to create limitations for navigation in the smaller bodies of water and parts of the high seas were getting under the various maritime territories of the littoral states and all of these had raised the concern among the non-littoral states of the enclosed or semi enclosed seas that the latter create obstacles for navigation of others under the pretext of special rights or circumstances for the enclosed or semi enclosed seas.

The representative of Iran during the earlier sessions of the 3rd UNCLOS had said:

" ...the issues of the enclosed or semi enclosed seas regarding management of the resources, international navigation and protection of the environment require special circumstances that are exception to the general rules of the law of sea...the freedom of navigation should be specific to the littoral states the enclosed or semi enclosed seas and for others, some other regime should be considered...passage of the non-littoral states from the strait of these seas should be only for proceeding to one of the ports of the littoral states. The enclosed or semi enclosed seas such as the Persian Gulf are considered as a destination, not a place of transit." (28)

Here we are seeing an effort to distinguish between the "sea of destination" with the "sea of transit". (29) in 1974, a series of rules were proposed for shipping in the enclosed or semi enclosed seas by Iraq and Algeria. (30) The Algerian draft referred to the free navigation of commercial vessels and state vessels that are used for commercial purposes in order to go to the littoral states of the enclosed or semi enclosed seas that a narrow outlet connected two parts of the high seas and traditionally used for international navigation. The Iraqi draft proposal referred to freedom of transit for all ships of the littoral states from the straits of the enclosed or semi enclosed seas. Iraq insisted that the freedom for passage should be safeguarded even if application of 12mile rule puts the entire waters of the straits in the territorial sea of the littoral states.

Of course the reason behind this argument was clear: Iraq did not want either Iran or Oman (the two states bordering the Strait of Hormuz) to exert much control on the shipping of the littoral states of the Persian Gulf as an enclosed or semi enclosed sea.

Although part of the working documents of the Caracas Conference was devoted to navigation in the enclosed or semi enclosed seas (31) and this issue was part of the discussions of the Geneva Conference of 1985, but no rule on this subject was included in the single negotiating text, which was the basis for the 1982 Convention. In the final sessions, Finland and Iraq made proposals for inclusion of certain rules in the convention regarding interruption in the navigation for the ships passing through the straits of the enclosed or semi enclosed seas due to construction if coastal installations, and artificial islands in the continental shelves of the littoral states of such regions. At the same period, UAE, Iraq and Yugoslavia proposed certain rules for guaranteeing freedom of navigation in the straits of the enclosed or semi enclosed seas but they never found their ways to the 1982 Convention.

During the fifth session of the 3rd UNCLOS, there were discussions about navigation in the enclosed or semi enclosed seas and some states wanted to include the following rules in the single negotiating text:

"1- in the straits used for international navigation, that connect the enclosed or semi enclosed seas to the other seas, there will be freedom of passage and flight over the strait without any hindrance for all ships and aircraft.

2- the above mentioned point will not affect the rules governing straits used for international navigation as set in the section two of this chapter." (32)

This proposal received many endorsements, but some others doubted its usefulness and eventually in the single negotiating text made no changes in the existing rules on the enclosed or semi enclosed seas.

Determination of maritime territories

Maritime territories here refer to the territorial sea, adjacent region and the EEZ or EFZ. Some states proposed that in the enclosed or semi enclosed seas, special rules should be used for delimitation of the maritime territories of the littoral states. Iraq and Thailand claimed that due to the disadvantaged geographical situation, a criterion other than median line should be used for this purpose and Iran supported median line for it. (33) Turkey proposed a set of rules regarding the islands in which it was stated that the maritime territories of the states bordering the enclosed or semi enclosed seas due to the special geographical circumstances should be determined jointly by the regional states. (34) Iraq in another occasion (the sixth session of the 3rd UNCLOS), proposed that delimitation of maritime territories of the states facing each other or situated next of each other in the enclosed or semi enclosed seas should be done with attention to the special circumstances and the principle of justice.

However, the 1982 Convention did not have any reference to the delimitation of maritime territories in the enclosed or semi enclosed seas.

Iran and the issue of the enclosed or semi enclosed seas

AS it was mentioned in various stages of the analysis of this issue, the fact that Persian Gulf is an enclosed or semi enclosed sea and the argument that implementation of the general rules of the international law of the sea in the Persian Gulf must be in a special way, were followed actively by the representatives of Iran in the sessions of the 3rd UNCLOS and even before that, in the 1958 conference. However, gradually it became clear that this concept is not moving in the direction that Iran wanted and the Iranian delegations in the final sessions of the 3rd UNCLOS dropped the idea. (35) There were different approaches to the idea of the enclosed or semi enclosed seas among the littoral states of the

Persian Gulf. Iran and Oman dominated the Hormuz Strait and they wanted to be able to control the traffic of non-littoral states especially the warships to and from the region. The other littoral states of the Persian Gulf were concerned that the states dominating the strait, may use their powers to control the shipping activities of other states in the Persian Gulf and they seriously supported the idea of freedom of navigation in the strait. At the same time, these states followed the idea of enclosed or semi enclosed seas for other issues (other than shipping). The approach of these states, especially Iraq that was entangled in a kind rivalry with Iran, was exactly compatible with the approach of the major maritime states as long as the shipping in the strait of the enclosed or semi enclosed seas was concerned. They wanted more freedom in such waterways. Iraq, more than any other state in the region, supported the idea of "transit passage" in the international straits. (36) the differences in Iran and Iraq's views can be easily traced by looking at the different draft proposals submitted by Iran (37) and Iraq (38) to various sessions of the 3rd UNCLOS regarding enclosed or semi enclosed seas.

The main reason of Iran for pursuing the idea of the enclosed or semi enclosed seas during the earlier sessions of the 3rd UNCLOS (and before that) was obtaining special advantages for the littoral states of the enclosed or semi enclosed seas, especially in the field of shipping and traffic of the other states warships. Given the position of Iran in the Strait of Hormuz this could lead to more control on traffic of ships to and from the Persian Gulf. This demand was not fulfilled during the 3rd UNCLOS and it reached no conclusion. (39). on the other side, along with failure of Iran in pursuing the idea of enclosed or semi enclosed seas the way it wanted, other developments happened that changed the entire problem in other directions. The most important case was development of the idea of Exclusive Economic Zone to the width of 200 miles from the baseline of the territorial sea. This idea became so strong during the sessions of the 3rd UNCLOS that many states in the world did not even wait for the final conclusion of the convention and they declared 200 miles (or less it was not possible) as EEZ and actually this practice turned into a customary part of the international law. The 1982 UNCLOS mentioned the concept in details. According to this concept, exploitation of the entire living and non-living resources in the waters and seabed of the EEZ belonged to the coastal states.

In a place like the Persian Gulf, the entire areas out of the territorial seas of the littoral states is EEZ and noting that their EEZs cannot be extended to 200 miles, they need to delaminate their parts. On these circumstances, Iraq has tried to use the idea of enclosed or semi enclosed seas for supporting its stance that due to the unfavorable geographical conditions of Iraq in the Persian Gulf, some other criterion should be used for delimitation. Also, during the war of Iran and Iraq (1980-1988), the government of Iran closed the Strait of Hormuz to Iraqi ships and those of others which carried contraband items to Iraq (40).

Conclusion

Given the above mentioned points, it can be concluded that the main objective of Iran in pursuing the idea of the enclosed or semi enclosed seas was not eventually included in the 1982 UNCLOS. On the other side, development of other concepts such as the Exclusive Economic Zone, answered the other objectives (other than shipping and regime of the strait) including prevention from access of non-littoral fishing ships to the living resources, and inclusion of such concepts in the convention, left no reason for Iran to follow the idea of enclosed or semi enclosed seas in the Persian Gulf. For the same reason, the representatives of Iran made no reference to this idea in the final sessions of the 3rd UNCLOS.

The duties mentioned in the article 123 of the 1982 Convention for the cooperation of the littoral states of the enclosed or semi enclosed seas are weak and completely voluntarily and non-obligatory. Therefore, it can be said that there is no special regime in the Persian Gulf (due to the fact that it is an enclosed or semi enclosed sea) according to the international law of the sea. The living resources in the Persian Gulf belong to the littoral states. The areas out of the national sovereignty and jurisdiction are free for international navigation. There is no special regime in the Persian Gulf, due to the idea of the enclosed or semi enclosed seas, for the shipping in the Strait of Hormuz or in order to delimit the maritime territories of the littoral states of the Persian Gulf.

Notes and sources:

(1) S. H. Amin, International and Legal Problems of the Gulf, London: Menas Press Ltd, 1981. P.16

(2) Iran has signed the 1982 UN Convention in the same year, but it has been ratified in Iran. Iraq, Kuwait, and Bahrain have signed and ratified this convention.

(3) The first UN Conference on the law of the sea in 1958 led to the conclusion of four conventions regarding territorial sea and adjacent region, continental shelf, fishing and living resources, and the high seas. The representatives of Iran were present in some of the sessions of the conference and the Iranian government is not member of any of those conventions.

(4) UN Doc. 1958 UNCLOS, Official Records, Vol.VI, 8th meeting (12 March 1958) p.14, Para 23.

(5) UN Doc. A/CON.13/C.2.L26,IV, Official Records. P.123

(6) Elizabeth Young and Vikta Sebek, "Red Seas and Blue Seas: Soviet View of Ocean Law", Survival, Nov/Dec. 1978, p.257

(7) B. Vukus, enclosed and semi-enclosed sea, International relations, Nos.11-12 (Tehran : Center for International studies, Tehran University, Spring 1978), p.171.

(8) Juridical Regime of Historical Waters, including Bays, International Law Commission Yearbook, 1962, Vol.2, pp.1-26

(9) R. R. Churchill and A. V. Lowe, The Law of the Sea, Manchester University Press, 1985, UK, p.30.

(10) F.I. Kezhevmikov, Kurs Mezhdunarondnoge Proavo. Moscow. 1972, Vukus, op.cit, p.175

(11) Ibid, p.176

(12) Vukus, op.cit. p.176

(13) Milan, Legal principles in the Persian Gulf, (in Persian), Persian Gulf and the Imposed War Seminar, Hormozghan Provionce, Mehr 1366 [September 1987] , p.1

(14) UNCLOS. Official Records, Vol.1,I ,pp.116,130,133,136,140,148,151,168.

(15) A/CONF.62/C, SEA/C/60 (17n July 1974), p.1

(16) A/CONF.62/C. 2/SR 43 (27 August 1974) p.10

(17) Press release, SEA/C/60, ((17n July 1974), p.1

(18) Vukas, op.cit, p.179

(19) A/CON.62?c.21.89,Rev.1. paras 8 and 17

(20) Vukas, op.cit, 183

(21) UNCLOS Official records, Vol. II, p.272

(22) Ibid. P.277

(23) Ibid., p.273

(24) Vukas, op.cit., p.187

(25) Ibid, p.188

(26) Ibid, p.277

(27) Ibid,p.273

(28) Vukas, op.cit.,p.187

(29) Ibid,p.188

(30) Press Release, SEA/C/112, UNCLOS, Caracas,38th Meeting, Second Committee, 13 August 1974, p.2

(31) United Nations Convention on the Law of the Sea, UN, NY, 1983, p.39

(32) A/CONF-62/C.2/SR.38, 15 August 1974, p.2

(33) A/CONF. 62/C.21/L.20 and L.21.
(34) Charles G. McDonald, Iran, Saudi Arabia, and the Law of the Sea, London, Greenweed Press 1980, p.184
(35) A/CONF.62/C. 2/W p.1. Provision 227, Formula A-C, Provision 228
(36) Vukas, op.cit. p.192
(37) UNCLOS, Official Records, VOL III. Pp.273-275
(38) A/CONF/62/C.2/L.55, Article 5
(39) Transit passage from the international straits was the idea supported by the major naval states and they entered it in the 1982 Convention instead of the innocent passage.
(40) UN Doc. A/CONF.62/C.2/L.72, 21 August 1974 Before the final sessions of the 3rd UNCLOS, there was a possibility of establishment of some kind of arrangements for the traffic of the non-littoral states in the Persian Gulf. Refer to: Charles G. McDonald, Iran, Saudi Arabia, and the Law of the Sea, London, Greenweed Press 1980 (he has said: creation of a special regime and closing the Persian Gulf to "unregulated passage" was emerging.

Iran and Persian Gulf Continental Shelf

According to the international law of the Sea, the outer boundaries of the continental shelf can extended many miles from the coastal state, however, Iran and other Persian Gulf states do not have the situation that countries bordering open seas and oceans have. The entire seabed of the Persian Gulf is continental shelf and due to the rich sources it has to be carefully delimited. Some sources believe that places like the Persian Gulf and the North Sea do not have continental shelf and their issues should placed under a different kind of heading such the seabed of the shallow waters.

Delimitation of the Persian Gulf between the littoral states has never been an easy task. For the same reason, even now some parts like the Iran-Iraq and Iran-Kuwait continental shelf (and therefore other maritime territories) are not delimited. There are many small and big islands, and sea elevations in the entire Persian Gulf that have added to the difficulties of the delimitation. Most of these sea elevations, and even islands were not of much importance in the history and many of them were left unclaimed and without any population.

Some of them were pinpointed in the old geographical maps only due to navigational requirements and safety of shipping.

Some of them also only were used in the past as bases for fishing. However, the discovery and importance of the oil and gas resources has changed the situation. Every piece of the Persian Gulf has turned into a precious place.

Many claims have appeared and even a small presence of a particular population has become important for these claims. In fact, noting the nature of the regional disputes in the Persian Gulf, which are almost entirely related to the oil and gas resources, many disputes first start between the oil and gas companies that try to do exploration and exploitation activities, and then they are pursued by the states.

Noting that according to the 1958 Geneva Convention and also 1982 UNCLOS the islands and even sea elevations (that stay out of water in the low tide), have certain territorial rights, the importance of the islands and sea elevations have been increased seriously.

According to the Article 15 of its Act on the Maritime Areas of the Islamic Republic of Iran in the Persian Gulf and the Oman Sea (1993), Iran's continental shelf beyond the territorial sea is as follows: "The provisions of article 14 shall apply mutatis mutandis to the sovereign rights and jurisdiction of the Islamic Republic of Iran in its continental shelf, which comprises the seabed and subsoil of the marine areas that extend beyond the territorial sea throughout the natural prolongation of the land territory."

Also, regarding the delimitation, the article 19 of the same law (please refer to the annex to see the full text of the Iranian law), the limits of the Iranian continental shelf unless otherwise determined in accordance with bilateral agreements, shall be a line every point of which is equidistant from the nearest point on the baselines of two States.

Iran has delimitation agreements with the Saudi Arabia signed in October 1968. The only problem that had made the delimitation of the continental shelf of the two countries delayed was the situation of Farsi and Arabi Islands and when the two sides agreed to give the Farsi to Iran and Arabi to the Saudi Arabia the rest was not so difficult.

Iran made the delimitation agreement with Qatar on 1969 and with Bahrain on 1971 and Oman in 1974. Iran has differences with Iraq which stopped it from delimitation. In fact, the Iraqis have made problems for conclusion of the delimitation agreement of Iran with Kuwait too.

Legal status of the Artificial Islands in the Persian Gulf

Persian Gulf is famous for oil, energy security and its geopolitics importance. However, the wealthy states of the Persian Gulf, especially the smaller states of the region, are adding another point of international attention to the existing ones: Artificial Islands. This is a growing feature in the region and noting the extent of various projects of the littoral states for establishment of the artificial islands it is going to become more important in future. The artificial islands are appearing one after another in the Persian Gulf. Some of the states in the Persian Gulf area are planning to expand their territories by establishing cluster of artificial islands. What we intend to do here is to explore:

1- What are the legal implications of the artificial islands in various maritime areas of the Persian Gulf?
2- What are the extents of the responsibility of the "flag sate" to the other states in a semi enclosed sea?
3- What are the environmental impacts of the artificial islands in the Persian Gulf and who is responsible for the damages?

Dubai is a forerunner in this field and many other small and at the same time rich states in the Persian Gulf may start going in the same direction. Iran has plans for establishment of artificial islands. (1) But the artificial islands of Kuwait, Oman, Qatar, and Bahrain are more important.

Although the issue of establishment of the artificial Islands has a longer history than the recent activities of Dubai (and a few other Emiratis inside the UAE, especially Sharejeh), the extent of the projects undertaken by Dubai (2) has opened the way for serious questions in a sensitive area. In the case of environmental pollution, there is a possibility of interpreting or trying to interpret the pollution as a form of international crime or even "aggression". (3) This has given rise to many legal questions such as the definition of the artificial Islands, legal status of the artificial islands in the maritime territories, the effects of the artificial Islands on the responsibilities of the coastal states in the sensitive area of the Persian Gulf and the effects of establishment of artificial islands on the rights of the other countries in one of the most important hot spots of the world. If you add the existing conflict of Iran and UAE on the three Islands of Tunbs and Abu-Musa, along with other problems of the region, to the picture, a more disturbing image may appear.

Maritime territories and the artificial islands in the Persian Gulf

According to the international law of the Sea (as a branch of the public international law), the territorial sea (or coastal sea) of the states surrounding the Persian Gulf (Iran, Saudi Arabia, Iraq, the United Arab Emirates or the UAE, Kuwait, Qatar, Bahrain, Iraq, and Oman) is 12 nautical miles, and outside of that there is a Contagious Zone of the same width and the rest is the EEZ or Exclusive Economic Zone. In other words, there is no "Res Communis" or part of the High Sea that is not a part of the EEZ in the Persian Gulf. This means that: no other state, except than the coastal states, can establish artificial islands in the Persian Gulf.

At the same time the geographical situation of the sea-bed in the Persian Gulf is in a way that the whole area under the waters of the Persian Gulf is part of the continental shelf and there is no area in the Persian Gulf out of the range of the legal or geographical range of the continental shelf. This requires that all surrounding states of the Persian Gulf delimit their shares of the continental shelf and the EEZ above it. This means: except than navigation over the EEZ, and the right of innocent passage from the territorial seas, there are no other rights (such as exploration and exploitation of resources, fishing, laying cables and so on) for the non-littoral states in the Persian Gulf. The maritime area behind the baseline (usually the coastal line) of the states is "internal waters".

Some of the states in the Persian Gulf area have very intensive straight baselines that are used for measuring their maritime territories which makes considerable parts of the coastal areas as internal waters. There is not any right for other states in this area. The location of the artificial islands in the various parts of the maritime territories may have different legal effects.

What is an artificial Island?

Artificial islands and installations are man-made, surrounded by water from all sides, above the water at high tide, supposed to stay at a specific geographical location for certain span of time and stationary in their normal mode of operation at sea. (4)

According to the law of the sea in general and the 1982 United Nations Convention on the Law of the Sea (UNCLOS) in particular, artificial islands have little legal implications, especially as far as the measurement of the maritime zones are concerned. The Artificial Islands are not considered as permanent harbor works. This means that: the coastal state can not claim the same rights that have been provided for the permanent harbor works in the determination of the baseline and measurement of maritime zones.

Article 11 of the UNCLOS 1982, provides: "for the purpose of delimiting the territorial sea, the outmost permanent harbor works which form an integral part of the harbor system, are regarded as forming part of the coast. Off-shore installations and artificial islands shall not be considered as permanent harbor works." (5)

According to the same convention the artificial islands are under jurisdiction a coastal state if they are constructed in the EEZ (the Persian Gulf has no place out of the EEZ). The article 56 of the 1982 UNCLOS provides: "1- in the exclusive economic zone, the coastal State has (a) Sovereign rights for the purpose of exploring and exploiting, conserving and managing the natural resources, whether living or non-living, of the waters superjacent to the sea-bed and the sea-bed and its subsoil, and with regard to other activities for economic exploitation and exploration of the zone, such as the production of energy from the water, currents and winds; (b) jurisdiction as provided for in the relevant provisions of this convention with regard to (i) the establishment and use of artificial islands, installations and structures....(2) In exercising its rights and performing its duties under this convention in the exclusive economic zone, the coastal state shall have due regard to the rights and duties of other states and shall act in a manner compatible with the provisions of this convention...." (6)

Artificial islands do not possess and are not able to gain the legal status of the natural islands or even the tidal elevations (i.e. Islands that only appear in the low tide). This means: they do not have their own maritime zones, such as territorial sea (12 nautical miles), contiguous zone (another 12 nautical mile) or exclusive economic zone (200 miles from the baseline).

In order to examine the legal basis of this point, let us have a look at a paragraph from the well-known book of Churchill and Lowe on the Law of the Sea, published by the Manchester University Press. The book has been translated to many languages and it has turned into a standard academic text book on the law of the sea and sometimes it is called the Bible of this branch. The authors have pointed out: "The only provision on artificial islands in the 1958 Geneva Convention (on the Law of the seas) was article 5(4) of the Continental Shelf Convention, which provided that installations connected with the exploration and exploitation of the shelf's natural resources and located on the continental shelf 'do not possess the status of islands. They have no territorial sea of their own, and their presence does not affect the delimitation of the territorial sea of the coastal states.' So the artificial islands cannot be the baseline for measuring the maritime territories. The 1982 Convention on the law of the sea supports this conclusion. Article 11 provides that: offshore installations and artificial islands shall not be considered as permanent harbor works…Article 60(8) and 80 provide that artificial islands and installations constructed in the EEZ or the continental shelf have no territorial sea of their own nor does their presence affect the delimitation of the territorial sea, EEZ, or continental shelf. …even though the construction of artificial islands on the high seas is now considered as a freedom of the high sea (article 87 of the Law of the Sea Convention), the prohibition on states from subjecting any part of the high seas to their sovereignty (article 89 of the 1982 Convention of the Law of the Sea) prevents the establishment of any maritime zones around artificial islands on the high seas." (7) As it was said the latter part has nothing to do with the existing situation in the Persian Gulf, because there is no part of the High Seas outside the EEZ in the Persian Gulf.

Only the coastal state may authorize the construction of artificial islands. This is a point expressed clearly in the article 60 of the 1982 UNCLOS. However, on the high seas beyond national jurisdiction, any state may construct artificial islands (article 87). (8)

Is it legally correct to say that any state in the Persian Gulf can construct artificial islands in its territorial sea without paying attention to the concerns of other coastal and even non-coastal states? The answer is definitely negative. Construction of artificial islands has considerable impacts around them, especially if the projects are large scales. The big projects, as the one Dubai is following, have extensive environmental implications for the whole region. "This subject is addressed in numerous treaties, other instruments, and international court and arbitral decisions. Late in 2003, for example, the International Tribunal for the Law of the Sea addressed claims by Malaysia that a Singapore land reclamation project would adversely affect the marine environment by, among other things, increasing sedimentation, increasing coastal erosion, and increasing salinity and pollution due to discharges. The Tribunal focused on a variety of procedural obligations, found in the Law of the Sea Convention and in general international law, which require steps of assessment, notification, and cooperation. For example, potential effects of activities reasonably believed to cause significant and harmful changes to the marine environment must be assessed, and other states must be notified about damage to the environment. The Tribunal in the Malaysia-Singapore Land Reclamation Case also invoked what it termed the "fundamental" duty to address environmental concerns through cooperation. In addition, the Tribunal found that, "given the possible implications of land reclamation on the marine environment, prudence and caution require[d]" the establishment of certain risk-assessment mechanisms. The Tribunal's "prudence and caution" Language indirectly echoed the still-evolving "precautionary principle." This principle has been articulated in treaty law and discussed in the international legal literature with increasing frequency since its inclusion in the Rio Principles, which were developed at the 1994 U.N. Conference on Environment and Development (UNCED). Broadly speaking, this principle provides that "[w]here there are threats of serious . . . damage, lack of full scientific certainty shall not be used as a reason for postponing cost-effective measures to prevent environmental degradation." Among other broad principles of current significance in the field of international environmental law, the concept of "sustainable development," articulated at UNCED, provides a procedural touchstone against which development proposals will be considered. An artificial island would have effects on the marine environment, effects that would be evaluated in light of these established and asserted principles of international environmental law." (9)

Persian Gulf is an enclosed or semi-enclosed sea. (10) Noting that the Persian Gulf is considered one of the "enclosed and semi-enclosed seas" what are the effects of this issue on construction of the artificial islands? Articles 122 of the UNCLOS, 1982, provides: "For the purposes of this Convention, "enclosed or semi-enclosed sea" means a gulf, basin or sea surrounded by two or more States and connected to another sea or the ocean by a narrow outlet or consisting entirely or primarily of the territorial seas and exclusive economic zones of two or more coastal States." This definition is exactly compatible with the Persian Gulf. As far the legal implications of this situation are concerned, article 123 of the UNCLOS 1982 provides: "State bordering an enclosed or semi-enclosed sea should cooperate with each other in the exercise of their rights and in the performance of their duties under this Convention. To this end they shall endeavor, directly or through an appropriate regional organization:

(a) to coordinate the management, conservation, exploration and exploitation of the living resources of the sea;
(b) to coordinate the implementation of their rights and duties with respect to the protection and preservation of the marine environment;
(c) to coordinate their scientific research policies and undertake where appropriate joint programs of scientific research in the area;
(d) to invite, as appropriate, other interested States or international organizations to cooperate with them in furtherance of the provisions of this article." (11)

Dubai (in the UAE) is home to some of the largest island complexes in the world, including the three Pam Islands Projects, the World, and the Dubai Waterfront. (12) Iranian officials have shown some concern about the implications of the construction of artificial Islands by the UAE in the Persian Gulf in the large scale. An Iranian top environmentalist has said the Persian Gulf is a semi-landlocked sea with great sensitivities and stressed that existence of ecosystems like Herra forests have added to the ecological significance of the area. (13) According to other Iranian sources: "...the UAE plans to build more than 325 artificial islands by expending billions of dollars in the Persian Gulf waters....the UAE currently possesses 60 square kilometers of the Persian Gulf islands and according to Sultan Bin Salim, the project manager of the World Islands, the UAE's island area increase to 1200 kilometers. The construction of these artificial islands will not only entail environmental degradation, but would also change the geopolitics of the Persian Gulf. The Persian Gulf is presently home to many species and biodiversity. More than 500 fish, 15 shrimp, and five rare turtle species inhabit the region. Experts of international organizations have warned against the destruction of Dubai's sole coral coast and coastal nests of turtles in the first phase of this project. They have also warned about the change in the quality of the Persian Gulf waters. Director General of Iran's Department of Environment, Seyyed Mohammad Baqer Nabavi, said he has called on the officials of the Regional Convention for Cooperation to Protect and Improve Coastal Zones and Maritime Environment to examine whether the construction of manmade islands in the Persian Gulf is in compliance with the international environmental standards. Referring to the 'very sensitive' nature of the Persian Gulf, Nabavi said any interference in its ecosystem as a result of these activities will negatively impact the region's environment, Fars News Agency reported. In an open letter to the UN Chief Kofi Annan, a copy of which was faxed to Iran Daily, Seyyed Mohammad Jafar Sadat Mousavi, who heads the Majles Environmental Faction, urged him to address the environmental concerns related to the World Islands. Sadat-Mousavi pointed out that the project is a blatant violation of the international conventions of 1958 and 1982 as well as their provisions." (14)

This concern is not limited to the Iranian authorities. "These artificial islands ...are a growing concern for environmentalists due to their impact on the local marine ecology. Dubai should be concerned as well for the long-term viability of the plan: rising sea levels from global climate change could spell trouble for its audacious and ostentatious investments..." (15)

Conclusion

It seems illogical and out of the legal context to ask the states around the Persian Gulf to stop building the artificial Islands in their maritime areas. However, taking into consideration the general obligations of the states as members of the international community, the treaty obligations of the states, especially in field of the international law of the sea, and the environmental law these points must can be stated as the conclusion of the present piece:

1- The non-littoral states cannot construct artificial islands in any part of the Persian Gulf.

2- The flag state (or the coastal state) of the artificial islands must give proper notification to the other regional states regarding the construction and particulars of the artificial islands.

3- The flag state (or the coastal state) of the artificial islands must arrange for regional coordination, especially through the ROPME for the sharing the information, exchange of the experiences, and planning for emergencies and other considerations related to the establishment and operation of the artificial islands. The Regional Organization for Protection of the Marine Environment (ROPME) which is based in Kuwait has been established in 1979 to look after the pollution and other environmental problems of Persian Gulf. Iran, Bahrain, Kuwait, Oman, Qatar, Saudi Arabia and United Arab Emirates (UAE) are members of the ROPME. It seems that the Persian Gulf States can use the ROPME for settling some problems regarding the artificial islands, and they have not used this channel so far. One of the issues mentioned in the "Protocol for Protection of the Marine Environment against Pollution from Land-based Sources" dated 1990, reads as "wastes generated from coastal development activities which may have a significant impact on the marine environment." (16)

4- The flag state (or the coastal state) of the artificial islands must make plans and preparations for observing the rights of other littoral states, and non-littoral states, especially in case of the using the right of innocent passage from the maritime territories. This may involve planning of the traffic schemes and corridors.

5- Noting the environmental effects of the artificial islands, the flag state (or the coastal state) of the artificial islands must make plans and provide regulations to minimize the environmental effects of the construction of the artificial islands. The effects of the artificial islands on the regional ecosystem and the marine pollution due to accidental and operational reasons are among the most important points that should be addressed.

6- In order to compensate for the inevitable damage to the regional ecosystem and the environment, the flag state (or the coastal state) of the artificial islands must take actions, through coordination with other concerned states, to perform plans and implement arrangements for replacing the damaged parts or systems. This may include construction of additional structures, establishment of systems to help the natural habitat of the species are put to danger due to the construction of the artificial islands, and provide alternatives for the concerned issues.

7- The flag state (or the coastal state) of the artificial islands must work with the other regional states. This is not only based on the fact the Persian Gulf is "enclosed and semi-enclosed sea", but also because it is mentioned as one of the environmentally sensitive areas in the Law of the Sea Convention.

8- The last but not least point is that all littoral states of the Persian Gulf are Muslim countries and there is a strong legal and religious principle in Islam which accepted both by Sunnis and Shiites and that is the Principle of "La-Ezrar" or stopping from creation of loss and damages to the interest of the other Muslims. This principle should be respected inn establishment of the artificial islands, even if it not rooted in the present body of the international law, as it is accepted in the Islamic law.

Notes and references

(1) Iran has apparently its own plans for the artificial islands but not in the Persian Gulf, but in the Caspian Sea. According to ISNA, the project of designing for the "Islamic republic of Iran Artificial island in Golestan Province (north of Iran), has been opened for bidding and this will be the first artificial island of Iran" http://www.isna.ir/main/NewsView.aspx?ID=News-695315&Lang=E, dated 04/11/2006.

(2) Regarding the various plans for establishment of the artificial islands in nthe Persian Gulf and their detailed maps and illustrations please see: Salahuddin, Bayyinah, the Environmental Impacts of the Artificial Islands Construction in Dubai of the UAE, Duke University, 2006, The Nicholas School of the Environment and Earth Sciences of the Duke University.

(3) Strategic Insights, Militarization of Energy Security, Vol. II, Feb. 2008, Daniel Moran and James A Russell, Center for Contemporary Conflict, Naval Postgraduate School in Monterey California, p.1.

(4) http://2100.org/w oceancitieslegal.html. The Legal issues of the Ocean Cities, By Cordula Fitz-patrick, Paris 1998, p.3

(5) The Law of the Sea, United Nations Convention on the Law of the Sea, United Nations, New York 1983.

(6) The Law of the Sea, United Nations Convention on the Law of the Sea, United Nations, New York 1983, P. 18

(7) R.R. Churchill and A. V. Lowe, The Law of the Sea, Melland Schill Studies in International Law, 3rd Edition, Manchester University Press, 1999, page. 51. This book has been translated into Persian by Bahman Aghai, and published by Ghanjedanesh Publications in Iran.

(8) http://www.answers.com/main.ntquery?tname=artificial%2Disland&print=true , Artificial Island.

(9) http://www.rwu.edu/macrocenter/2004/confer-encepapers/noyes.htm, New Land for peace. An overview of International Legal Aspects, by John E. Noyes, 2004.

(10) Definition and description of the semi-enclosed seas are mentioned in this article. However, it is necessary to point out that the Third United Nations Conference on the Law of the Sea that led to the conclusion of the UNCLOS 1982, or Montego Bay Convention, could not divide the definition for enclosed and semi-enclosed seas.

(11) The Law of the Sea, United Nations Convention on the Law of the Sea, United Nations, New York 1983.

(12) http://www.answers.com/main.ntquery?tname=artificial%2Disland&print=true , Artificial Island.

(13) http://wwwirna.ir/index2.php?option=comnews&task=print&code=0608056833200220, dated August 5, 2006.

(14) http://www.iran-daily.com/1385/2624/html/national.htm, dated 30th of July 2006
(15) http://news.mongabay.com/2005/0823-tina_butler_dubai.html, dated 23 August 2005.
(16) http://www.ropme.com/pages/legal.htm.

Legal Aspects of Marine
Pollution in the Persian Gulf

Persian Gulf is in the south of Iran and Iran has the longest shores in the Persian Gulf than any other country. Persian Gulf is an important and sensitive region because of its oil and gas resources and the ways to transport them. It is surrounded by oil and gas producing and exporting countries (OPEC and OAPEC) which supply 60 percent of the oil needed in the West. Exploration, exploitation, transportation of energy by ships and pipelines, are only some of the factors that create environmental dangers in the Persian Gulf. However s certain other factors are contributing to this point:

1- The high volume of traffic of the oil tankers
2- The limited capacity of the region: waters of the Persian Gulf cover an area of some 241000 square kilometers. (The entire seabed in the Persian Gulf is continental shelf of the adjacent states and these are no areas out of the EEZ or exclusive economic zone).

3- The Persian Gulf is generally a shallow body of water and its average depth is only 40 meters and its deepest parts are only 100 meters.

4- Connection of the Persian Gulf to the Sea of Oman (and the Indian Ocean) is through the narrow outlet of Hormuz Strait and this stops the Persian Gulf from easy interaction with the high seas out of it and therefore restricts its ability to mix its waters with the outer currents

5- Persian Gulf is the loading point for the oil tankers and these large ships always fill up their tanks with water (called ballast water) in order to keep their balance in the journeys. They discharge the diluted water of the tanks in the loading points (this is called operational or normal pollution caused by the oil transportation).

6- The possibility of accidents for the tankers and oil and gas installations (in the case of tankers accidents, this is called accidental or unintentional pollution)

7- The hostilities in the region and military operations involving attacks against oil and gas installations and routes (the case at hand was setting fire to the Kuwaiti oil wells by the Iraqis in the course of their invasion). In fact, the international law has not much about this kind of pollution and whatever is there is related to outdated wars.

Iran has always insisted on the special characteristics of the Persian Gulf and necessity of special regulations or even a special regime for control of the pollution in the region. In 1973 London Conference on Prevention of Oil Pollution from Ships, Iran tried and convinced the conference to declare the Persian Gulf as one of the "special zones".

Also, Iran is an active member of the 1978 Kuwait Convention for Combat against Pollution of the Persian Gulf. The members of this convention have established RAPME, as a center to confront oil spills, discharge of oil from wells and oil tankers. The general rule is that the states living on the coats of a common body of water should not damage the rights of others in the process of their use of the body of water.

1958 and 1982 conventions require every state to draw up regulations and take measures to cooperate with the competent international organizations to prevent pollution of the seas. The 1982 Convention obliges the sates to cooperate in the implementation of international rules relating to liability and compensation and asks states to consider measures to protect rare and fragile ecosystems.

There are two conventions which establish strict but limited liability for damages from oil spills. They focus especially on the consequences of the damage incurred. These are the 1969 International Convention on Civil Liability for Oil Pollution Damage and the 1971 Convention on the Establishment of an International Fund for Compensation for Oil Pollution Damage. But the liability rule of these two conventions applies to the pollution damage caused by vessels and not by states. Actually, Article XI of the 1969 Convention points out that the provisions of this Convention shall not apply to warships owned by a state and used for non-commercial purposes. (1)

"In order to protect the marine environment of the Persian Gulf and the Sea of Oman against oil pollution, the Kuwait Convention and its Protocols were developed by the countries of the region in cooperation with the United Nations Environment Program (UNEP). In accordance with Article 16 of the "Convention", Regional Organization for the Protection of the Marine Environment (ROPME) was established in 1978 (Kuwait) and based on Article 3 of the "Protocol on Regional Cooperation in Combat of Marine Emergencies", Marine Emergency Mutual Aid Center (MEMAC) was established in 1983 (Bahrain)." (2)

Notes

(1) http://auhf.ankara.edu.tr/dergiler/auhfd-arsiv/
AUHF-1991-1992-42-01-04/AUHF-1991-1992-42-01-
04-Turgut.pdf

(2) http://www.sid.ir/en/ViewPaper.
asp?ID=129146&varStr=8;FARSHCHI%20
P.,DABIRI%20FARHAD,SHOJAEI%20
SARA;JOURNAL%20OF%20ENVIRON-
MENTAL%20SCIENCE%20AND%20
TECHNOLOGY;SUMMER%202008;10;2%-
20%2837%29;76;86

The Greater and Lesser Tunbs and Abu-Musa

The Greater and Lesser Tunbs and Abu-Musa Islands are situated near the Straight of Hurmuz in the Persian Gulf, south of Iran. The Lesser Tunb is 22 miles from the mainland of Iran. The Lesser Tunb is 17 miles from the Iranian land. Both of them are not able to sustain living and they had never inhabitants. Abu-Musa is the home for a limited number of people (less than 50 households). The Greater and Lesser Tunbs and Abu-Musa Islands have been part of Iran since the times immemorial. In the Nineteenth Century, they were parts of the "Lengheh Territory" that was itself an administrative section of the Fars Province of Iran.

Sovereignty of Iran over these Islands have been recorded in many books, historical documents, almanacs, maritime journals, geographical maps (that show the three concerned islands in the color of the Iranian mainland), official documents, administrative reports, the officials notes of the British authorities in India and so on. The British authorities created some difficulties for the Iranian governments in the case of the Iranian control of these Islands during the early Twentieth Century. The actions of the

British officials were always facing protests by the Iranian local and state authorities.

In 1968 the British decided to withdraw from the East of Sues by 1971. The British made a package deal with Iran according to which Iran stopped its demand for restoration of its sovereignty over Bahrain, and take back its three islands of The Greater and Lesser Tunbs and Abu-Musa Islands. Only in the case of Abu-Musa, Iran accepted to give some advantages to the Sheikdom of Share-jeh. The Iranian government accepted this formula only in the hope of supporting the small states of the Persian Gulf and giving them a chance of getting independence. The British authorities were the only officials that Iran made the arrangements with them because at that time the states like the UAE were not established yet. The Shah of Iran faced a great difficulty in making the people of Iran ready for such arrangement and some political groups never accepted that.

It is interesting that following the restoration of the Iranian sovereignty over the three islands, three Arab countries complained against Iran in the United Nations Security Council. The United Arab Emirates (which had been formed of several Sheikhdoms with the support of Iran) was not one of them. Egypt, Iraq and Libya were the parties to the dispute. They claimed that Iran has occupied part of the Arab lands. The reason was that Egypt under the control of Jamal Abdul-Nasser (the fabricator of the name of the Arabian Gulf), and Iraq, and Libya were thinking that they were the main leaders of the Arab world and they were pretending that they acted as the representative of the Arabs. However, the UNSC heard the explanations of the parties and after hearing the report of the British representative that implicitly referred to the "package deal", the UNSC deleted the issue from its agenda.

It is also interesting that Iraq, under Saddam Hussein, after attacking Iran in 1980 and capturing a part of the Iranian territory in the border of Iran-and Iraq declared that if Iran wanted peace, it must accept several conditions, including the withdrawal of Iranian forces from the three islands of the Greater and Lesser Tunbs and Abu-Musa. (The most important part of the other conditions was abrogation of the 1975 treaty between Iran and Iraq which designated the Thalweg line or the most navigable canal in the Arvandrood as the border of the two countries).

Saddam dropped the condition when the Iranian forces pushed the Iraqis back and entered the Iraqi territory. Since then the Arab States at Persian Gulf have repeatedly claimed that Iran should hand over the three islands to them. They have succeeded to get the backing of the Arab League. Also, during the last several years they have recruited some of the top research institutes and legal experts in Western world for finding grounds for their claims in the international law and politics.

The British are playing a two-sided and mostly anti-Iranian game in these regards. They have already cast doubt on one of the most important sources of Iran's claim by saying that the old map of the British Authorities in India (the map that was formally presented to Iranian officials as the expression of the maritime situation in the region and it showed the three concerned islands in the color of mainland Iran) was not official. Also, they are not ready to give clear explanations about the "package deal" that resulted in Iran's withdrawal of claims over Bahrain and get its sovereignty back in the three islands.

In fact the story of Bahrain's independence was a clear indication of the package deal. For the same reason the UAE's advocates have always tried to deny the existence of the package deal between Iran and the British authorities that led to the independence of Bahrain, UAE and restoration of Iranian rights in the Persian Gulf over the Tunbs and Abu-Musa.

Following the package deal in 1968, the case of Bahrain was put in the agenda of the United Nations. The United Nations chose a representative on the issue of Bahrain. He traveled to Bahrain and talked to several people in streets and later reported to the UN that the people of Bahrain wanted inexpedience. There was no referendum, public inquiry, research work or even a simple random sampling in the standard model. It was not clear that how the UN representative had reached such a conviction that the people of Bahrain wanted independence, while more than half of the people of Bahrain were Iranians and they wished to remain Iranian. However, due to the fact that the "package deal" was there, the UN did not go through such questions and accepted that Bahrain should be independent. Following this development the representatives of the Western states, especially the British officials, thanked Iran for the peace loving actions and understanding the international situation.

Inspection of ships by the naval forces in the maritime territories

There is no right in the international in general and law of the seas in particular for the naval units of one state to stop and inspect the ships of others states in the high seas. In the territorial sea, contiguous zone and the EEZ (Exclusive economic zone, which the coastal state has certain rights, the coastal states may ask the ships of other states that it involved in the activities which are under the jurisdiction of the coastal state, to leave those areas. Articles 89 and 90 of the United Nations 1982 Law of the Sea Convention have prohibited the states from exerting their control on other states in the high seas. (1) Although Iran and the US are not parities to the concerned convention (2) this part of the convention is reflection of the customary international law and both states are subject to it.

Also, only on the conditions of war (when the state of war is established), and when the international law of the wartime is observed, the naval units of the parties to the war (the hostile parties) can inspect the ships of third parties if they suspect that they are carrying contraband items to the destination of the other party to the war. This right is not limited to any place in the high seas (for example places close to the international straits which are good places usually for enforcing such rights) and can be enforced in the high seas anywhere.

However, when the "state of war" is not established and the international law of the wartime is not in force, one of the only ways that inspecting ships of other countries can be done is a res- olution by the Security Council of the United Nations. Such a resolution can be the basis for such rights by all members of the United Nations against a target state. "Most states currently believe that only a UN resolution can authorize state inspections on the high sea" (3) this right is not limited to any state and all mem- bers of the United Nations Organization (and in one interpreta- tion, even those that are not members of the UNO) can choose to enforce this resolution.

Therefore, the resolutions 1803 and 1929 of the UNSC that have been adopted in the framework of the UN measures to stop the nuclear program of Iran from moving towards military objectives, and declared certain sanctions against Iran due to the nuclear program of Iran, constitutes such a base for the naval units of the all member states of the UN to inspect the ships going to and coming from Iran to make sure that they are not carrying the prohibited items.

The government of Iran has called the 1929 Resolution of the UNSC as worthless piece of paper and the Iranian military (4) and civilian officials (5) have declared that if other states try to implement the parts of the 1929 resolution of the UNSC that is related to the inspection of ships, it would consider such action as a hostile move and take action forcefully against the members of the UN that exert their rights.

The speaker of the Iranian parliament has said: "we warn the US and certain adventurist countries that if they are tempted to inspect the Iranian air and ship cargos, we will take tough actions against their ships in the Persian Gulf and the Sea of Oman." (6) And also, Commander of the Islamic Revolutionary Guards Corps

Naval Forces, Ali Fadavi, has said: "IRGC will take retaliatory measures in case Iran-bound cargo ships come under inspection by the West." (7) It is clear that the announcement of the Iranian regime is against the international law and the Iranian government has no right to do the actions that it has declared.

It is obvious that Iran's claims have no basis in the international law, but it is based on the political strategy of Iran which says: the area is either safe for all not safe for all. This is not a new policy and in fact during the last years of the Iran-Iraq war (1980-1988) that the "Tanker Wars" had started, Iran was following the same policy.

The issue did not go further than attacking the oil tankers of other countries (which resulted in the escorting of Kuwaiti oil takers by the US Forces in the region and later to a clash between the Iranian and US naval forces that resulted in destruction of almost half of the Iranian navy), but this time it may turn into a total war by Iranian attacks to the main oil terminals and trying to block the Strait f Hormuz.

That is why many sources believe the passage of the even the former UNSC resolution of sanctions (1803) against Iran, makes a war between the United States and Iran more likely. (8) the article 11 of this resolution provided:

"11. *Calls upon* all States, in accordance with their national legal authorities and legislation and consistent with international law, in particular the law of the sea and relevant international civil aviation agreements, to inspect the cargoes to and from Iran, of aircraft and vessels, at their airports and seaports, owned or operated by Iran Air Cargo and Islamic Republic of Iran Shipping Line, provided there are reasonable grounds to believe that the aircraft or vessel is transporting goods prohibited under this resolution or resolution 1737 (2006) or resolution 1747 (2007)" (9)

Therefore the concerned part of the 1929 was only a reaffirmation of what was already adopted by the UNSC. The 1929 resolution provides that:

"14. Calls upon all States to inspect, in accordance with their national authorities and legislation and consistent with international law, in particular the law of the sea and relevant international civil aviation agreements, all cargo to and from Iran, in their territory, including seaports and airports, if the State concerned has information that provides reasonable grounds to believe the cargo contains items the supply, sale, transfer, or export of which is prohibited by paragraphs 3, 4 or 7 of resolution 1737 (2006), paragraph 5 of resolution 1747 (2007), paragraph 8 of resolution 1803 (2008) or paragraphs 8 or 9 of this resolution, for the purpose of ensuring strict implementation of those provisions" (10)

The USC resolutions have opened the way for the international community and especially the West and the US to take action against the regime of Iran if they wish to do so. In this context, William Hawkins says:

"The resolution was passed under Chapter 7 of the UN Charter dealing with "Threats to the Peace, Breaches of the Peace, and Acts of Aggression" and which discusses proper responses that include the use of force. The resolution will be used as an excuse by those who do not wish to act, but a bold administration could also use its language against Iran as justification for taking decisive action if it wanted to do so. The question the world is waiting to hear is which interpretation President Obama will embrace." (11)

Notes:

> (1) Article 89 of the UNLOSC: "Invalidity of claims of sovereignty over the high seas no State may validly purport to subject any part of the high seas to its sovereignty."And the Article 90 "Right of navigation every State, whether coastal or land-locked, has the right to sail ships flying its flag on the high seas."
>
> (2) Iran has signed the 1982 UN Convention on the Law of the Sea and has not ratified it.

(3) Legal Basis of the Naval Interdiction under the proliferation security initiative, by Sue Soo-ha Yang, Brooklyn Law School, 1 Oct. 2003.

(4) Iran's parliament adopts bill against inspections, http://www.google.com/hostednews/ap/artcle/ALeqM5iRqiZV1Meppj40thTs81BOv4Dds, dated 07/20/2010 .

(5) Inspection of Iranian cargo ships imperils regional security, July 2, 2010, http://www.imra.org.il/story.php3. Iranian Defense Minister ....Iranian officials say that UN Security Council resolution 1929 which calls for restrictions on the country's shipping industry as well as goods coming to and leaving Iran is in violation of international law.

(6) Ibid

(7) Ibid.

(8) United Nations Security Council Resolution1803 : casus belli, http://www.globalreserach.ca/index.php?context=va&aid=8308

(9) http://www.iaea.org/NewsCenter/Focus/Iae-aIran/unsc_res1803-2008.pdf

(10) http://www.america.gov/st/texttrans-english/2010/June/20100609150613ptellivremos0.8968622.html

(11) http://www.aim.org/guest-column/un-resolution-on-iran-open-to-interpretation

Iran and Landlocked states

The UN Convention of 1982 (article 69) has been devoted to the right of landlocked states to participate in exploitation of the surplus of living resources in the EEZ. The other articles of the convention that are related to the landlocked states are articles 61, 62, and 297. These articles provide that the landlocked states can participate in a suitable manner in exploitation of the surplus resources in the EEZ of the same "region".

It is necessary to point out that according to the convention, the coastal states should determine their capacity for exploitation of the living resources in the EEZ and they can define a level of surplus for exploitation of the landlocked states. The regulations for this type of exploitation should be agreed by bilateral or multi-lateral agreement, of the concerned states.

The AALCO (formerly AAL CC) has conducted a study due to the request of the Nepal on this subject, and it has reviewed possible agreements in such cases between the landlocked state and the transit or coastal state.

In the Convention of 1982, there are 9 articles (124 to 132) in the chapter 10 for the "right of access of the landlocked states to the sea and freedom of transit for them."

Article 124 has defined the landlocked states:

"1. For the purposes of this Convention:

(a) "land-locked State" means a State which has no sea-coast;

(b) "transit State" means a State, with or without a sea-coast, situated between a land-locked State and the sea, through whose territory traffic in transit passes;

(c) "traffic in transit" means transit of persons, baggage, goods and means of transport across the territory of one or more transit States, when the passage across such territory, with or without trans-shipment, warehousing, breaking bulk or change in the mode of transport, is only a portion of a complete journey which begins or terminates within the territory of the land-locked State;

(d) "means of transport" means:

(i) railway rolling stock, sea, lake and river craft and road vehicles;

(ii) where local conditions so require, porters and pack animals.

2. Land-locked States and transit States may, by agreement between them, include as means of transport pipelines and gas lines and means of transport other than those included in paragraph 1." Article 125, has reiterated the right of access and described the manner the right can be enforced.

"*Article 125*

Right of access to and from the sea and freedom of transit

1. Land-locked States shall have the right of access to and from the sea for the purpose of exercising the rights provided for in this Convention including those relating to the freedom of the high seas and the common heritage of mankind. To this end, land-locked States shall enjoy freedom of transit through the territory of transit States by all means of transport.

2. The terms and modalities for exercising freedom of transit shall be agreed between the land-locked States and tran-

sit States concerned through bilateral, sub-regional or regional agreements.

3. Transit States, in the exercise of their full sovereignty over their territory, shall have the right to take all measures necessary to ensure that the rights and facilities provided for in this Part for land-locked States shall in no way infringe their legitimate interests."

However, the conditions for passage of the landlocked states from the transit state for access to the sea is left for agreement of the two sides or regional and sub-regional agreements. Part one of the Article 125 of the 1982 Convention does not make the right itself dependent on those kinds of agreements. It simply recognizes the right. At the same time, it seems this is not an absolute right and it should be considered along with paragraphs 2 and 3 of the same article.

In fact, the 1982 UN Convention has left the landlocked states in confusion. Article 125 provides that "reasonable" regulations should be used for arranging the freedom of transit of the landlocked states, but it does not specify which those regulations are. At the same time, the convention does not put a commitment on the transit states to refrain from creating constraints for landlocked states. The paragraph 3 of the same article gives complete rights to the transit states to take all measures necessary to ensure that the rights and facilities provided for in this Part for land-locked States shall in no way infringe their legitimate interests.

This means that passage of the land-locked States should not harm the laws and regulations of the transit state and should not be against its security and financial interests. Therefore, the transit state can take necessary measures to prevent the smuggling activities through its territory and transportation of arms. The questions that: what else can be considered as the Legitimate Right or is it possible to stop passage at certain occasions totally , or as far as certain types of items are concerned, are not clear.

In addition to the 1982 Convention, there are other documents about the land-locked States. Among them are Versailles Treaty, GATT (1947), Havana Charter (1984), 1958 Convention on the High Seas, and 1965 Convention on transit trade of the land-locked States. Also, we should refer to the Resolution 10282 (11) and Resolution 1015 (11) of the UNGA on 20 and 21 February 1957, the 1981 UN Convention on the least developed countries, UNCTAD, TD/B/AC.39/3 dated September 1985 of the experts for transit passing of the land-locked States.

The 1965 Convention (New York Convention) on Transit Trade of the land-locked States has emphasized on the agreement between the two sides. At the same time, it has not made the right itself dependent on others and only the method of enforcement is what should be determined by the agreement.

At the moment, there are 42 land-locked States (see the list at the end of this section) and out these, a considerable number are in the area that Iran is situated: Afghanistan, Azerbaijan, Kazakhstan, Tajikistan, Turkmenistan, and Uzbekistan. Unlike the European land-locked States that have practically solved their problems with the transit states, the issue is still subject to many problems in the Middle East and Central Asia regions. The situation in the European region has benefited from the modern communications, mutual trade ties and geographical characteristics of the concerned countries which are in the heart of the Europe. The geographical situation of the African and Asian land-locked States is not similar to the European ones and for the same reason; the European pattern is not a formula that can be used in these cases. At the same time, the most important point in the agreements between the European countries and the land-locked States is the "principle of reciprocity" in them.

There are also bilateral agreements for facilitating the passage of land-locked States through the transit states. Among them we can refer to: Agreement of Russia and Afghanistan dated 18 June 1955, and Agreement of Nepal and India dated 31 June 1950.

The official view of Iran on this, based on the statement of Iran on the signing of the 1982 UN Convention on the Law of The Sea (by 2010, Iran has not yet approved the convention, but it has not made any indication regarding detrainment from its approval and it is committed to it in the framework of the legal obligations of states in the distance between signing and approval of a treaty) is that: passage is not an absolute right and it is a kind of a concession to the land-locked States and therefore, the coastal or transit state can demand something in return for facilities that it provides to the land-locked States. The access of the land-locked States to the sea is based on the entire parts of the article 125. The transit state can take all measures for ensuring that its sovereign rights are not violated through the transit passage of the land-locked States and these calls for conclusion of specific agreements between the two sides and the agreements should be based on reciprocity.

List of Landlocked Countries of the World:
Afghanistan
Andorra
Armenia
Austria
Azerbaijan
Belarus
Bhutan
Bolivia
Botswana
Burkina Faso
Burundi
Central African Republic
Chad
Czech Republic
Ethiopia
Hungry
Kazakhstan
Kosovo

Kyrgyzstan
Laos
Lesotho
Liechtenstein
Luxembourg
Macedonia
Malawi
Mali
Moldavia
Mongolia
Nepal
Niger
Paraguay
Rwanda
San Marino
Serbia
Slovakia
South Ossetia
Swaziland
Switzerland
Tajikistan
Turkmenistan
Uganda
Uzbekistan
Zambia
Zimbabwe

Articles 124 to 132 of the UN 1982 Convention on the Law of the Sea on the Landlocked States

Article 124

Use of terms

1. For the purposes of this Convention:

(a) "land-locked State" means a State which has no sea-coast;

(b) "transit State" means a State, with or without a sea-coast, situated between a land-locked State and the sea, through whose territory traffic in transit passes;

(c) "traffic in transit" means transit of persons, baggage, goods and means of transport across the territory of one or more transit States, when the passage across such territory, with or without trans-shipment, warehousing, breaking bulk or change in the mode of transport, is only a portion of a complete journey which begins or terminates within the territory of the land-locked State;

(d) "means of transport" means:

(i) railway rolling stock, sea, lake and river craft and road vehicles;

(ii) where local conditions so require, porters and pack animals.

2. Land-locked States and transit States may, by agreement between them, include as means of transport pipelines and gas lines and means of transport other than those included in paragraph 1.

Article 125

Right of access to and from the sea and freedom of transit

1. Land-locked States shall have the right of access to and from the sea for the purpose of exercising the rights provided for in this Convention including those relating to the freedom of the high seas and the common heritage of mankind. To this end, land-locked States shall enjoy freedom of transit through the territory of transit States by all means of transport.
2. The terms and modalities for exercising freedom of transit shall be agreed between the land-locked States and transit States concerned through bilateral, sub-regional or regional agreements.
3. Transit States, in the exercise of their full sovereignty over their territory, shall have the right to take all measures necessary to ensure that the rights and facilities provided for in this Part for land-locked States shall in no way infringe their legitimate interests.

Article 126

Exclusion of application of the most-favored-nation clause application of the most-favored-nation clause

The provisions of this Convention, as well as special agreements relating to the exercise of the right of access to and from the sea, establishing rights and facilities on account of the special geographical position of land-locked States, are excluded from the application of the most-favored-nation clause.

Article 127

Customs duties, taxes and other charges

1. Traffic in transit shall not be subject to any customs duties, taxes or other charges except charges levied for specific services rendered in connection with such traffic.
2. Means of transport in transit and other facilities provided for and used by land-locked States shall not be subject to taxes or charges higher than those levied for the use of means of transport of the transit State.

Article 128

Free zones and other customs facilities

For the convenience of traffic in transit, free zones or other customs facilities may be provided at the ports of entry and exit in the transit States, by agreement between those States and the land-locked States.

Article 129

Cooperation in the construction and improvement of means of transport

Where there are no means of transport in transit States to give effect to the freedom of transit or where the existing means, including the port installations and equipment, are inadequate in any respect, the transit States and land-locked States concerned may cooperate in constructing or improving them.

Article 130

Measures to avoid or eliminate delays or other difficulties of a technical nature in traffic in transit

1. Transit States shall take all appropriate measures to avoid delays or other difficulties of a technical nature in traffic in transit.
2. Should such delays or difficulties occur, the competent authorities of the transit States and land-locked States concerned shall cooperate towards their expeditious elimination.

Article 131

Equal treatment in maritime ports

Ships flying the flag of land-locked States shall enjoy treatment equal to that accorded to other foreign ships in maritime ports.

Article 132

Grant of greater transit facilities

This Convention does not entail in any way the withdrawal of transit facilities which are greater than those provided for in this Convention and which are agreed between States Parties to this Convention or granted by a State Party. This Convention also does not preclude such grant of greater facilities in the future.

National Interests of Iran in the Caspian Sea

The Caspian Sea littoral states have failed to reach a general compromise on the legal regime of the Caspian Sea. Since the collapse of the USSR, these states have convened many conferences in all levels, including the first summit in 2002 in Ashgabat (Turkmenistan) and second summit in Tehran, 16th of October 2007, to solve this problem and they have not succeeded. This issue has the potential to turn into a point of confrontation and even conflict, especially with discovery of oil and gas resources and the new importance of the Caspian oil as an alternative to the Persian Gulf oil (at least to some extent).

It seems that under the present conditions, the best policy for the Islamic Republic of Iran is refraining from entering into any kind of treaty for the legal regime of the Caspian Sea, because the conditions are set to impose the worst situation upon Iran. Iran has no reason to hurry about the legal regime of the Caspian Sea.

What are those positions?

Although the USSR is dead and the Russian Federation is not a super power as it was once, the Russian leaders are always dreaming of restoring the Russian hegemony in the area that once used to be the Russian domain. As far as the Caspian Sea is concerned, they want to use the whole Caspian Sea for their military and civilian fleet. They are following these policies:

1- Division of the Caspian Sea bed (only) on the basis of a modified median line (MML). It means the more coastal area you have, the more area of the Caspian Sea you get. According to the MML, Russia, and Azerbaijan get almost twenty percent (each of them), Kazakhstan gets 30 percent, Turkmenistan gets almost 17 percent and Iran gets almost 13 percent of the Caspian Sea-bed. The MML formula leaves the waters of the Caspian Sea for common use of the littoral states.

2- Putting pressure on all Caspian states, especially Iran, to accept the MML for division of the Caspian Seabed. The Russians have succeeded to convince Azerbaijan and Kazakhstan in this field. Iran, along with the on and off support of Turkmenistan, has not agreed with it.

3- Excluding all non-littoral states from having military or civilian presence in the Caspian Sea. The Russians have stationed one of their most important naval concentrations in the Caspian Sea. The civilian fleet of the Russians in the Caspian Sea handles ninety percent of the maritime transportations in the Caspian Sea. The Fishing fleet of the Russians has no rival in the region. They want to exclude the non-littoral states to have no rival. The other littoral states have nothing considerable in the Caspian Sea, except than some old dated boats and the fishermen who work in the way the ancient tribes.

4- Creating difficulties for the usage of Volga-Don and Volga-Baltic channel for the other littoral and non-littoral states, for keeping the advantages of the Russian fleet, ports and facilities. The Russians have been insisting that the Volga channel is completely an internal waterway. (While the new conditions of the Caspian Sea requires some kind of reconsideration in this regard and make it an international waterway or a waterway under a special regime, such as the Bosporus and Dardanelle.)

5- Refraining from providing the other littoral states with larger ships for expansion their military or civilian fleet. For example by refraining from selling ships, or helping them to build naval facilities. The littoral states of the Caspian Sea, except than the Russian Federation, do not have any military of civilian fleet (Iran's share from the shipping in the Caspian Sea is less than 4 percent.) and the Russians want to keep them that way.

6- Forcing the littoral states to use Russian outlets for the export of their oil and gas. The landlocked states of the Caspian Sea need proper outlets for their exports and the Russians try to make them use the Russians facilities. One of the ways to do so is the rejection of building under water pipelines in the Caspian Sea under the pretext that it damages the environment. It is noteworthy that the Russians are responsible for ninety percent of the pollution in the Caspian Sea through thousands of the Russian factories that pour their industrial wastes in the Volga River and eventually the Caspian Sea.

7- Formation of a kind of common military force for the Caspian Sea. This force will be almost completely a Russian instrument for patrolling all the Caspian Sea. Other littoral states have hardly enough boats to do low-level police activity in their shorelines.

8- Using the opportunity gained by Iran's isolation to force Iran to accept the MML. Iran is under pressure and the Iranian regime is desperate for its survival. The Russians are well aware that they cannot treat a thoroughly nationalist government in Iran, as they treat the Islamic regime of Iran.

The Republic of Azerbaijan is happy to get twenty percent of the Caspian Sea by the MML. However their policies are:

1- Attracting the Western countries, especially the USA into the Caspian Sea. The inclination of Azerbaijan to the Western states, especially the USA, is not originating from an inherent love. This policy is based on the fact that the Azerbaijan Republic, as the second Shiite country in the world (after Iran), is feeling worried about the ideological provocations orchestrated by the Islamic Republic and other Islamic extremist elements. Also, the Azeris need to neutralize the Russian presence, as a force supporting Armenia (which has close relations with Iran and Russian Federation).

2- Good relations with Israel as an indication of the inclination to the Western countries. The Azerbaijan Republic is aware that its relations with Israel can play an important role in convincing the West about its intentions.

3- Presenting the Baku-Jeyhan pipeline as the best way for oil exports of the Caspian land-locked countries. The Baku-Jeyhan pipeline is now operational and despite the fact that it was not an economical project, the Western support has succeeded to create this pipeline. The Baku-Jeyhan pipeline is the clear sign of the failure of Iran and Russia in the regional pipeline diplomacy. However, the Azeri oil is not enough for using the full capacity of the Baku-Jeyhan pipeline, and Azerbaijan needs to attract the cooperation of the regional countries, especially Kazakhstan to give this pipeline.

4- Promoting the proposed Gas Pipeline called Nabucco, as an alternative route for carrying gas from the Caspian area, cutting the monopoly of the Russians in the regional business and creating a "gas" Baku-jeyhan.

5- Getting into NATO and leaving the hand of NATO free in the Caspian Sea. Azerbaijan has already suggested the

Americans and the NATO to use the Abshoran peninsula as their military bases. There are some news about
establishment of the radar posts by the NATO in Azerbaijan and possible use of the Azeri territory for an attack
against the Islamic regime of Iran.

6- Getting the international support in the case of Nagorno
Gharabagh with Armenia. This is the most important
issue in the political agenda of the Azeri governments.
Azerbaijan is ready to give concessions in the Caspian
Sea to the forces that help it in the case of Nagorno
Karabagh. The Republic of Azerbaijan has rejected the
suggestions of Iran for meddling in this issue because
they do not believe in impartiality of Tehran.

7- Exploration and exploitation of the resource in the Caspian Sea with the capital and expertise of the Western
countries. Azerbaijan has been exploring the oil resources
of the Caspian Sea for the last two hundred years (more
seriously in the last fifty years). They need new technology and investment in the oil and gas resources.

8- The Azerbaijan Republic has already joined with the
Russians in using the MML as the formula for division
of the maritime borders with the Russian Federation, as
far as the Caspian Seabed is concerned. However, they
are interested to make this division wider and to include
the waters too.

Kazakhstan is trying to make use of the opportunity created by
the access of the country to most of the Caspian Sea. The MML
leaves this country with 30% of the Caspian Sea-bed. The Kazakhstan's fields are actively developed by the Western companies,
especially the Americans, interested in non-OPEC, non-Arab,
Non-Iranian oil. Kazakhstan has already concluded treaties with
the Russians and the Azerbaijan Republic for using the MML as
the division criteria of the Caspian Seabed. Iran has proclaimed
such treaties as null and void because the littoral states have origi-

nally agreed to make decision on the legal regime of the Caspian Sea unanimously.

The government of Turkmenistan is not satisfied with the MML, not because its share according to the MML formula is 17 percent, but due to the fact that the important oil fields claimed by Turkmenistan are given to Azerbaijan by the MML. Turkmenistan once went to the brink of war with Azerbaijan over these oil fields (Kapaz or Sardar oil fields). It was interested to be in the side of Iran against the MML, but it was not ready to tie its destiny to the Islamic regime of Iran. Turkmenistan has already showed that it agrees with the MML and there are only some problems (such as the Kapaz oil fields) that should be hammered out. Also Turkmenistan is waiting for the destiny of Turkmenistan-Afghanistan-Pakistan pipeline.

What is the position of Iran?

The position of Iran is to divide the Caspian Sea according to equity (20% for each). The Russian Federation is imposing its formula of MML for the division of the Caspian Sea-bed and leaving the superjacent waters for the common use. Iranian position about the possible division of the Caspian Sea is not limited to the "seabed" (unlike MML). Iran is asking for a complete division of the whole sea. This kind of division will lead to:

1- Restriction of the Russian forces from traveling freely all over the Caspian Sea.
2- Stopping the industrialized fishing fleet of the Russians from using the national sections of the other countries
3- Disconnection the direct link of the Russians with Iran. The Russian Federation has no land border with Iran at the moment. Following the collapse of the USSR in 1991, the land border of Iran and the Russians was removed. Division of the Caspian Sea into national sectors, as Iran

is calling for, will result into removing the water borders with the Russians too.

However, at the moment nobody is taking the positions of Iran in the Caspian Sea serious. Due to the troublesome nature of the Iran's Islamic regime, all countries in the region are pausing to see what is going to happen to the regime of Iran. After all, what is the use of entering into agreement with an unstable regime? The successors of this regime may decide to punish or take revenge from those who support it now. Iranian people think that the failure of the Islamic regime of Iran for protecting the Iranian rights in the Caspian sea (as an ancient country which has been living in the southern shores of the Caspian Sea and as a state that has shared this body of water with the Russians for a long time) is the result of the mismanagement of the international relations and the wrong decisions of the Islamic regime in the field of the foreign policy.

The best policy for the Islamic Republic of Iran, as a regime that has not succeeded to safeguard the national interests of Iran in the Caspian Sea, is refraining from entering into any kind of contractual arrangements or agreements that might jeopardize the national interests of Iran in the Caspian Sea in an irreversible way. These are the reasons for the preferred inaction policy:

1- Iran does not need its oil and gas resources in the Caspian Sea immediately. There are many places (including the Persian Gulf) that Iran possesses large amounts of oil and gas reserves. These can be exploited much easier as compared to the Iranian side of the Caspian Sea. In fact, the Iranian side of the Caspian Sea is deep (the deepest point is almost a thousand meters deep) and it is difficult to explore and exploit oil and gas reserves here. Any economical activity in this section requires high technology and more investment. It must be noted that although the Caspian Sea is a lake, it has the features of the open sea in many regards like water currents and weather conditions. You to add to this picture the difficulties of getting the facilities to the required points in the face of the non-cooperation of the littoral states and isolation of Iran.

2- The nuclear issue of Iran, along with other policies of Iran, has left Iran in a weak situation in front of the Russians. Iran needs the Russians for stopping the adoption and implementation of the UN sanctions. This makes the maneuverability of the already weak policy of Iran more limited.

3- It is not a good idea to take the case of the Iranian interests in the Caspian Sea to the international tribunals (such as the UNSC, International Court of Justice, and international arbitrations). Iran has not the international prestige, the support of any country in the world and the case of the Caspian Sea the Russians are on the other side too. Referring the case of the Iranian rights to the international forums will not be a solution for Iran at the present conditions. In fact, Iran must try to avoid the efforts of the others to take the case to such forums. With all littoral states, Russia, the US and the EU on the other side, who is going to vote for the Iranian rights in the Caspian Sea?

4- The establishment of the new legal regime will ease the way for the others to do what they want and Iran will be left back due to the lack of expertise and financial resources. The picture is bleaker for Iran if we consider that some of the most important oil and gas fields in the Caspian Sea are common among the littoral states in any kind of division.

5- Iran has failed in the pipeline diplomacy so far. The important pipelines are already avoiding Iran. The new legal regime will not change the situation of Iran in the pipeline diplomacy of the region.

Caspian Sea: Potentials for Conflict

Leaders of the Caspian Sea littoral states constantly talk about peace and security in the Caspian region. They even signed a security agreement at the third summit of the Caspian states (Baku, Azerbaijan Republic, 18 November 2010). The reality, however, is that the Caspian Sea is not as peaceful as it seems. The region has serious potential for turning into a flashpoint for confrontation and conflict. Not only have the littoral countries of the Caspian Sea failed to solve their problems, they have actually taken steps to further militarize the region.

What are the problems?

The division of the Caspian Sea remains a thorny issue complicating relations among the littoral states. Since the collapse of the former USSR, these states made efforts (such as the Ashgabat summit in 2001 and the Tehran summit in 2007) to arrive at a collective solution; they failed. Thus, bilateral agreements among some littoral states have begun to overshadow efforts at collective diplomacy, resulting in the conclusion of several treaties among Russia, Azerbaijan, and Kazakhstan. The so-called southern states of the Caspian Sea, namely, Iran and Turkmenistan, have refused to go along, declaring these agreements null and void.

Nor has the conclusion of bilateral treaties among the northern Caspian states resolved all issues. The concerned treaties focus on the division of the seabed on the basis of the modified (equidistance or) median line (MML), leaving many other issues unresolved. The formula, devised by the Russians, leaves the waters of the Caspian Sea free for shipping of all littoral states (and does not clarify shipping by non-littoral states). Other than the Russians, the littoral states do not have important naval units or commercial ships in the Caspian Sea. So it is clear that the formula used in the concerned bilateral treaties serves the interests of Russia above the others. Furthermore, these agreements make no distinction between military and commercial shipping, leaving the door open for all sorts of disputes.

If the provisions of the international law of the sea regarding maritime areas are applied in the Caspian Sea (as suggested by some states, without requiring compliance by the littoral states), a number of issues will demand attention: territorial water, baselines, internal waters, river mouths, bays, ports, islands and their territories, low-tide elevations, innocent passage of commercial and military units, submarine traffic, passage through the Volga-Don waterway, sea lanes, traffic separation schemes, passage of nuclear powered ships, warships of the littoral and non-littoral state, responsibility of the flag state, hot pursuit, regulations governing safety of life at sea, certification of seaworthiness, indemnity for damages from shipping and pollution, contagious zone, research and survey activities, economic zones, regulations for laying pipelines, responsibility for accidental and operational oil and nuclear pollution, and so on.

Iran and Turkmenistan do not agree with the criteria used by others for the division of the seabed in the Caspian Sea. Iran insists that the division of the Caspian must be based on equitable and just principles, giving equal shares to all five states. Having failed to convince the others to accept the common administration of the sea, Iran is now insisting on equitable division of the entire Caspian. In 2004, Iran's representative in Caspian affairs, Mehdi Safari, said Iran had prepared documents explaining that, according to international law, Iran's share of the Caspian must be 20.4 percent (interview with Iranian TV, 10 April 2004), but the Iranians have never made these documents publicly available. Iran demands control of the Alborz/Alove oilfields, which Azerbaijan also claims, while Turkmenistan and Azerbaijan contend for ownership of oilfields (Sardar/Capaz) that both claim.

In the years that Iran insisted on "equity" in division of the Caspian Sea, it seemed this might imply acceptance of something less than 20 percent for all the littoral states (for example, 16 or 17 percent for Iran or any formula that included a couple of known oilfields such as Alborz—which Azeris call Alove or Flame). But before the 2010 Summit in Baku, Iranian officials made it clear this was not the case. Immediately after the conclusion of the meeting of the Caspian ministers in Tehran (15 November 2010), the special envoy of the Iranian president for Caspian Sea affairs, Mohammad Mehdi Akhundzadeh, responded to a question by the official news agency of Iran, IRNA, about a 20 percent share for Iran: "Our aim goes further than this limit."

Combining these regional issues with the existence of undemocratic, corrupt, and unstable governments in the littoral states of the Caspian Sea, the inclination of the great powers to use Caspian oil as a rival or alternative to OPEC and OAPEC oil, and the expansion of NATO toward the East, one sees the picture of oil, blood and politics that Alfred Nobel saw a century ago.

Militarization efforts

During the last couple of years, the militarization of the Caspian Sea has been a hot topic in all the meetings of the regional states at various levels, including the last session of the foreign ministers in Tehran. As early as 2000, the Stockholm International

Peace Research Institute (SIPRI) convened a conference entitled "National and Regional Security of the Central Asian States in the Caspian Sea Region" (22-23 September 2000, Almaty, Kazakhstan). The conference concluded, inter alia, that:

"The Central Asian region is now a zone of acute political instability and the national security of regional states is challenged by a wide variety of political, military and socio-economic threats, both from within the region as well as from the outside.... The main domestic threats are associated with the declining state of living standards of the majority of the local populations, as well as with growing inter-ethnic and inter-confessional tensions and conflicts, while the main external threats to national and regional security are posed by religious extremism supported from abroad, international terrorism as well as illegal trade in arms and drugs. Among other major threats and risks to national and regional security were the unresolved issues of the legal status of the Caspian Sea, problems of transportation of oil and gas to the world market, as well as territorial and border problems between regional states."

The Russian Federation has been the frontrunner in the militarization of the Caspian Sea. In fact, the Russians have indirectly used the demonstration of their military power to convince the others to accept the type of bilateral treaties that they prefer.

"The growing complexity of political-economic interests in the region has forced Russia to change its position on the Caspian's status on more than one occasion. Not only are the Caspian's resources at stake, but also transportation networks, commercial operations, the status of the Sea itself and the issue of military control over the region.... The military issue has developed into an area of intense concern of late.... Such an arrangement would greatly heighten tensions in the Caspian region and could lead to war.... The increased US and NATO attention toward the region prompted one Russian general to claim that the greatest threat to Russia is not China or Islamists but the possibility of Desert Storm II starting on the shores of the Caspian over economic issues."

(Timothy L. Thomas, Russian National Interests and the Caspian Sea, Perceptions, 1999-2000, vol. IV, no. 4, pp. 750-96.)

Of course, the Americans are there, too. Post-9/11 US strategies, especially the security of Israel, the war against terrorism and control of oil resources, have opened a new stage in the role of the USA in the region. Economically, the Americans are as interested as the Russians in Caspian oil. At the same time, from a military point of view the Americans are interested in moving their forces to places closer to hot spots (Afghanistan, Iraq, Iran, the Persian Gulf, and Caspian Sea). The new American military bases are likely to be small but equipped with rapid-response forces in the region. After this stage, many of the older types of bases, such as Incirlik in Turkey—which proved useless in the invasion of Iraq—will be closed.

Azerbaijan Republic is prime real estate for the American presence. In every possible way, the Azeris have been calling for the Americans to have a presence there. Unhappy with the US assistance to Azerbaijan, Tehran has complained that this represents a military build-up against itself. Iran is also locked in a dispute with Azerbaijan over ownership of an oil-rich corner of the Caspian Sea, resulting in a 2001 clash between an Iranian warship and an Azeri oil research vessel. A real threat from Iran to Azerbaijan may reveal Russia's role in this process in a new light. Russia will neither support Azerbaijan in the open, nor quarrel with Iran. Most likely it will play the role of a peacekeeper. Whatever the case, signs point disturbingly to contention, even bloodshed, for Caspian energy resources.

In an English-language article titled "War for Caspian Sea Inevitable?" the Russian newspaper Pravda writes: "Problems exist in the relations between Azerbaijan and Turkmenistan, as well as between Azerbaijan and Iran. These countries still argue about the borders of their sectors of the sea. The Caspian dispute has triggered the militarization of the Caspian Sea." (http://www.pittsreport.com/2010/11/war-for-caspian-sea-inevitable/)

In the Caucasian Review on International Affairs, Alexander Jackson, writes: "Kazakhstan is negotiating to buy corvettes armed with formidable Exocet anti-ship missiles (Eurasianet, 23 June 2010), whilst Russia's Caspian Flotilla is being boosted with new frigates (News.az, 2 November 2010), and Azerbaijan is strengthening its radar and command-and-control systems. Even Turkmenistan is trying to increase its naval profile (Jamestown Foundation, 16 February 2010). The white elephant in the room is Iran's naval aspirations, framed by its stubborn and isolated position on Caspian delimitation. Earlier this year, Iran announced the launch of its first destroyer in the Caspian, capable of electronic warfare, anti-submarine and anti-aircraft attacks—in short, far more firepower than necessary to stop sturgeon poachers (Trend.az, 19 February 2010). Moscow and Tehran are uneasy about each other's naval presence, but have so far presented a united front to prevent their biggest fear—a greater role in the Caspian for the US or NATO (Eurasianet, 19 November 2010). These fears presumably informed the security agreement which emerged from the Baku summit." (http://cria-online.org/CU - file - article - sid - 103. html) Of course, the same article makes a serious mistake about the positions of Iran about the legal regime of the Caspian Sea, assuming that Iran is retreating from its past positions.

Regarding the relations of Azerbaijan and Turkmenistan as far as the Caspian Sea in concerned, Anar Valiyev says: "In 1997 [Kazakh President Saparmurad] Niyazov accused Azerbaijan of illegally exploiting the Azeri and Chirag oilfields, and threatened to sue the oil companies involved. That same year, pressure by Turkmenistan caused Russian companies Rosneft and Lukoil to withdraw from a project to develop the Kyapaz oilfield. At the same time, a consortium of foreign oil companies led by the Bechtel Corporation proposed the construction of a Transcaspian gas pipeline to transport Turkmen gas to Turkey through Azerbaijan. In the absence of any significant gas reserves of its own, Azerbaijan's role was going to be one only of transit. Niyazov's death in 2006 heralded the start of a new era in Turkmen-Azerbaijani relations.

Many expected that the proposed Nabucco pipeline, designed to connect Caspian gas fields to Europe through Turkey, would provide a natural point of alliance and cooperation for Turkmenistan and Azerbaijan. However, differences between the countries persist…. Analyzing the foreign policies of Turkmenistan and Azerbaijan, it is easy to see that both countries are moving in different directions rather than approaching a common position…"
(http://www.gwu.edu/~ieresgwu/assets/docs/pepm_087.pdf)

Conclusion:

For the foreseeable future the Caspian Sea will remain a hot spot in international affairs. This importance results from regional and global geo-politics, oil resources, terrorism, and narco-trafficking. The littoral states of the Caspian Sea are inclined to gradually strengthen their military forces in the Caspian, triggering the era of militarization in the Caspian Sea. Such an increase in militarization makes a bad mixture with the regimes all around the Caspian Sea, deeply plunged as they are in undemocratic and unstable dictatorships, corruption, violations of human rights and fundamental freedoms, discrimination, nepotism, social gaps and ethnic rivalries.

Legal Aspects of Marine Pollution in the Caspian Sea

The Caspian Sea, the biggest lake in the world, is so polluted that it can be considered in a serious environmental danger. Five littoral states of the Caspian Sea (i.e. Iran, Russia, Turkmenistan, Azerbaijan and Kazakhstan), especially Russia, are adding to the pollution of the Caspian Sea in an enormous scale. United Press International reported in 2004 (1) "the Caspian Sea, the largest inland body of water on Earth, is in danger of turning into an environmental dead zone, a development whose impacts would be felt throughout Central Asia and East Europe."

The littoral states of the Caspian Sea have failed to find a new legal regime for the Caspian Sea, following the collapse of the former USSR.(2) This has increased the lack of regulations and legal measures for protection of the Caspian environment. Out of frustration from the final settlement of the legal regime for the entire Caspian Sea, the littoral states signed convention for the protection of the Caspian environment in 11/05/2003 (Turkmenistan signed it at a later date). However, this did not solve the problem. The convention was a regional instrument and the main tasks were left for the future protocols.

At this juncture (April 2008), no protocol is signed for this end yet. The convention only provided that the littoral states were committed to take steps individually or collectively to reduce and control the pollution. It does not make clear the responsibility of states if they fail to comply with their international commitments under numerous international instruments (the list is provided in this article) for prevention, creation, reduction, cleaning up and compensation for the damages of the pollution, especially for the marine pollution by oil (accidental, intentional, unintentional and operational).

Russians are the greatest polluters of the Caspian Sea. They create 80% of the Caspian pollution. After that, Azerbaijan is producing some of the worst kinds of pollutions because of their outdated oil refineries and other oil installations in the Caspian Sea. Kazakhstan and Turkmenistan are after Azerbaijan in the pollution production. Iran has the lowest share in the pollution of the Caspian Sea. According to the report of the Energy Information Administration in 2000: Untreated waste from the Volga River, into which half the population of Russia - and most of its heavy industry - drains its sewage, empties directly into the Caspian Sea. Oil extraction and refining complexes in Baku and Sumgayit in Azerbaijan are major sources of land-based pollution, and offshore oil fields, refineries, and petrochemical plants have generated large quantities of toxic waste, run-off, and oil spills. In addition, radioactive solid and liquid waste deposits near the Gurevskaya nuclear power plant in Kazakhstan are polluting the Caspian as well... The impact on human health has been measurable, and the Caspian's sturgeon catch has decreased dramatically in recent years, from 30,000 tons in 1985 to 13,300 tons in 1990 and then to as low as 2,100 tons in 1994. (3)

Local and international environmental groups point out that the Caspian ecosystem has already suffered decades of abuse from the Soviets, and is fragile and in need of recovery; not additional stress. Decades of lax environmental controls have dumped dangerous toxins into the Volga River, the main source of the Caspian and into the sea itself. Scientists estimate that each year an average of 60,000 metric tons of petroleum byproducts, 24,000 tons of sulfites, 400,000 tons of chlorine and 25,000 tons of chlorine are dumped into the sea. Concentrations of oil and phenols in the northern sea are four to six times higher than the maximum recommended standards. Around Baku, where oil drilling and industrialization have been happening for almost a century, these pollutants are ten to sixteen times higher (4)

The Caspian sturgeon and the Caspian seal, one of two freshwater species in the world, have been dying in large numbers as a result of polluters or poachers, who have operated with impunity since the collapse of the former Soviet Union. "The sturgeon will be commercially extinct in two to three years," says a World Bank official. (5) According to a report by AP, dated 21 June 2000, thousands of seals have died in Caspian Sea (6): "Thousands of dead seals have been found along Kazakhstan's Caspian Sea coast, in an outbreak that officials blame on unusually warm weather. But environmental experts say is connected to oil pollution. Workers have collected and destroyed the bodies of 11,000 dead seals."

Galina Yeroshenkova, an Emergency Situations Agency official, said." Problems of Caspian Sea pollution can be divided into three types which are:

1. Chemical pollution by the running rivers.
2. Ecological problems, connected to the rise of the level of water.
3. Offshore oil industry. (7)

The offshore oil industry in Azerbaijan sector of Caspian Sea has developed since 1949. On platforms "Neft dashlari" and "28 April" heavily developed productions and transportation of oil. At this time sulfuric oil for processing in refiners was transported to Baku from the Kazakhstan coast by tankers. As a result of oil flood during its production and transportation the level of sea pollution by oil exceeds allowable norm in some sites up to 20 times.

Among the most polluted sites of Caspian Sea by oil are: Baku bay, Apsheron archipelago, Islands, Turkmenbashi, Cheleken, Mangishlak, Tengiz and other sites of oil industry. (8)

An encouraging sign has been a move towards greater cooperation in protecting the Caspian. Several initiatives have boosted regional cooperation in protecting the environment, including the establishment of the Caspian Environment Programme (CEP) in conjunction with the Global Environmental Facility. The overall goal of the CEP is defined as "environmentally sustainable development and management of the Caspian environment, including living resources and water quality, so as to obtain the utmost long-term benefits for the human populations of the region, while protecting human health, ecological integrity, and the region's sustainability for future generations." (9)

I believe that the most important factor in endangerment of environment in the Caspian Sea is oil pollution and all other pollutions that come with it. Exploration and exploitation of oil and gas resources in the Caspian Sea is the main activity of the future all around the Caspian Sea. Therefore, a special attention to the oil pollution is necessary.

Happily, the body of laws and regulations concerning the oil pollutions very advanced in the international law. The littoral countries have to agree to apply most of those laws and regulations to the environment in the Caspian Sea. For this purpose, they do not even need to wait until the whole issue of the Caspian Sea's legal regime is solved because by that time, may be nothing much is left to take care of it. The most important instrument of the international law in this case is the 1973 London Convention (MARPOL) or the International Convention for the Prevention of Pollution from Ships. Generally the oil pollutions are the result of two main categories of factors: The operational or deliberate factors, and accidental or un-deliberate factors.

In the field of operational factors the "ballast water" has a special position. This is the water taken on by all kinds of ships, especially oil tankers when they are not carrying oil cargoes, to keep them operating smoothly. Naturally they throw the water, which is contaminated when they want to reload. This causes a considerable amount of pollution. 1973 Convention has very important articles for these cases.

The other documents that should be taken into consideration are:

1- Convention for Prevention of Marine Pollution by Dumping from Ships and Aircraft (Oslo 15 Feb.1972).
2- Convention relating to Civil Liability in the field of Maritime Carriage of Nuclear Material (1971).
3- International Convention on the Establishment of an International Fund for Compensation of Oil Pollution Damage (1971).
4- Contract regarding an Interim Supplement to Tanker Liability for oil Pollution (1971, CRISTAL).
5- International Convention for the Prevention of Pollution of the Sea by Oil (London 1954).
6- International Convention on Civil Liability for Oil Pollution Damage (1960).
7- Offshore Pollution Liability Agreement (OPOL).
8- Tanker Owners' Voluntary Agreement concerning Liability for Oil Pollution Damage (1969).
9- Agreement concerning Cooperation in Measures to deal with Pollution of the sea by Oil (1971).
10- Convention for the Prevention of Marine Pollution from Land-based sources (1974).
11- International Convention on Civil Liability for Oil Pollution Damage resulting from Exploration for, or Exploitation of, Submarine Mineral Resources (1977).
12- Agreement relating to the Establishment of Joint Pollution Contingency Plans for Spills of Oil and other Noxious Substances (1974).
13- Agreement of Cooperation regarding Pollution of the Marine Environment by Discharge of Hydrocarbons and other Hazardous Substances (1980).

14- Protocol of 1984 to amend the International Convention on the establishment of an International Fund for Compensation for Oil Pollution Damage1971 (1984).

15- Tanker Owners Voluntary Agreement concerning Liability for Oil Pollution (T0VALOP).

In addition to the above instruments, which have a direct relevance to the issue at hand, there are other sets of regional and secondary documents that must be taken into consideration for the purpose of defining an effective regime for the protection of the Caspian Sea environment against various pollutants, especially pollution by oil:

I. Kuwait Regional Convention for Cooperation on the Protection of the Marine environment from Pollution (978), along with its protocols for on cooperation in oil pollution emergencies, and land based pollutions.

II. Parts of 1982 UN convention on the Law of the Sea which are related to marine pollution.

III. Convention on the Protection of the Marine Environment of the Baltic Sea Area (1974).

IV. Convention for the Protection of the Mediterranean Sea against Pollution (1976) along with its protocols on Damping (1976), Cooperation in Emergencies (1976), Land-based sources of Pollution (1980) and Protected Area (1982).

V. Regional Convention for the Conservation of the Red Sea and Gulf of Aden Environment (1982).

VI. Convention for Cooperation in the Protection and Development of the Marine and Coastal Environment of the West and Central African Region (1981), along with its protocol for cooperation in emergencies.

VII. Agreement for Cooperation in dealing with Pollution of the North Sea by Oil (1969).

The future of the Caspian Sea depends on how successful will be the Caspian littoral states in finding of a suitable formulas out of all these documents for protection of the unique environment of the Caspian Sea. The most visible ways in this line are as follows:

1- The experience of the Kuwait regional Convention (1978), which has resulted in establishment of "Regional Organization for Protection of the Marine Environment" (ROPME) in the Persian Gulf, can be very useful in the case of the cooperation of concerned states in the Caspian Sea.

2- Most of general and important instruments regarding the protection of the marine environment, especially those related to the oil pollutions, have reached a stage that is called "mandatory." This means that all states have to be observe them, even if they are not directly party to them.

3- The Russian Federation is responsible for a considerable amount of pollution in the Caspian Sea. At the same time, it is signatory and party to almost all-important conventions in the law of the sea and pollution. The only point is that they do not consider themselves as committed to observe those obligations in the Caspian Sea. This approach has to change. The Republic of Azerbaijan, which is responsible of polluting the Caspian Sea by oil during the last 50 years or so, should accept the commitments to observe internationally recognized standards of prevention of oil pollution in these areas.

4- Out of the several thematic centers that are established by the CEP (Caspian Environment Programme) in the littoral states of the Caspian Sea, the Legal Center, which is in charge of preparing regulations, is in Moscow. I do not think that Russians are very interested in preparing regulations which most of them would address themselves. May be these centers should circulate among the concerned states, before becoming fully independent from the CEP.

5- The local oil and gas companies (like NIOC in Iran, SOCAR in the Republic of Azerbaijan, and Russian companies) should adopt special environment friendly policies in their activities in the Caspian Sea as opposed to their practices in the past and as a guideline for international oil companies. This is especially important because these companies themselves takes part in oil and gas exploration and exploitation activities in the other Caspian Sea states, in addition to what they do in their own states. Unfortunately, at the moment these local oil and gas companies are working under worst conditions in the Caspian Sea.

6- The international oil and gas companies, such as Mobil, Chevron and BP, which are active in the Caspian Sea area, and those which are planning to be present in the oil and gas scene of this area, should accept to observe the same standards of operations that they have in places like the North Sea and Gulf of Mexico, as far as the protection of the environment is concerned.

7- In preparing legal documents and operational standards in the Caspian Sea, due attention should be given to the current international regulations and standards regarding the special areas. In these areas (such as the Antarctic waters) in addition to the general rules and regulations designed to protect all environments, some particular regulations are in place because of the special geographical or physical characteristics of the areas. In the case of the Caspian Sea, the fact that this body of water is not really connected to the open seas of the world makes it imperative to have special rules and standards.

8- Establishment of "reception facilities" in certain parts of the Caspian Sea is a necessary action. As it was mentioned before, one the important sources of oil pollution in the marine environment is the ballast water. These reception facilities can receive water, which is mixed with oil residues, and return it to the sea after processing.

9- Establishment of oil and gas pipelines in the Caspian Sea should be subject to the internationally recognized standards for protection of the environment. There are many international documents that can be used as a guide, and these are some of them: The Agreement relating to the Transmission of Petroleum by Pipeline from Ekofisk Field and Neighboring Areas to the United Kingdom (1973), the Agreement relating to the Exploitation of the Frigg Field Reservoir, and Transmission of Gas there from to the United Kingdom (1976).

Nuclear pollution is one of the less known and rarely discussed dimensions of the serious pollution problems in the Caspian Sea. At the same time, the radioactive contamination is one of the most damaging and dangerous types of pollution in the world. The nuclear activities of the coastal states, implications of the former nuclear explosions, the remnants of the nuclear tests, the nuclear wastes (which will be radioactive for thousands of years) and finally the nuclear side of oil exploration and exploitation and transportation (especially by pipelines) are the sources of nuclear danger in the Caspian Sea. The nuclear pollution is not in the same level all over the Caspian Sea. It is different from the viewpoints of sources and types of dangers. But if we take into consideration that the Caspian Sea, as the greatest lake in the world, is not connected through natural channels to the high seas, and at the same time, a strange wind current (there are other strange characteristics in the Caspian Sea like its rising and falling levels that is still subject to controversy) keeps steering the water inside the lake like a giant spoon, then the general dangers of the nuclear pollution for all coastal states, even those who have smallest role in the nuclear contamination of the Caspian Sea (like Iran) become more evident. It is necessary that the countries of the Caspian Sea

region try to establish a system for combating the nuclear pollution in this area as soon as possible. The role of the atomic energy
organizations in each of these countries and also the International
Atomic Energy Agency in this field is very important. Fortunately,
at the moment, there are several major international instruments
that can be used by the concerned parties for protection of the
Caspian Sea against the nuclear pollution. I believe even before
the legal regime of the Caspian Sea is finally determined among
the coastal countries, serious steps must be taken in this field.
Some of the most important international instruments that must
be observed in the process are:

1- Vienna Convention on Civil liability for Nuclear Damage, 1963 and its 1997 Protocol.
2- London Convention on the Prevention of Marine Pollution by Dumping of Wastes and other Material, 1972.
3- Convention on Physical Protection of Nuclear Materials,
1980.
4- Monte go Bay (united Nations) Convention on the law
of the seas, 1982.
5- Basel Convention on the Trans-boundary Movement of
the Hazardous waster and their Disposal, 1989, and its
Protocol of 1999 on Liability and Compensation.
6- Vienna Convention on Nuclear Safety, 1994.
7- Vienna Convention on Supplementary Compensation
For nuclear damage, 1997.
8- Helsinki Convention on the Protection and Use of Trans-
boundary Waters and International Lakes, 1992, and its
Protocol of 1999.
9- Vienna Joint Convention on the safety of Spent fuel
Management and on the Safety of Radioactive Waste
Management, 1997. This Convention is especially important for transportation of nuclear materials taking into
consideration the recent arrangement between Iran and
the Russian Federation for sending the spent fuels of the

Bushehr Nuclear Power Plant and possibly other power plants in Iran. The negotiation for construction of 4 to 10 other nuclear power stations in Iran by the Russians are under way, according to the strategic program of cooperation of Iran and Russian Federation.

Notes

(1) Marina Kozlova, "Will Caspian Sea become another Aral?", United Press International, 28 June 2004 (http://www.eurocbc.org/caspiansea_may_suffer_same_fate_as_aralsea_28jun2004page1696)

(2) please see : Bahman Aghai Diba, "the Legal Regime of the Caspian Sea: any changes in the positions of Iran?", Soochow Law Journal, Vol. II, January 2005, Number one, pp. 239-252.

(3) http://www.eia.doe.gov/emeu/cabs/caspenv.html

(4) Rachel Neville, Environmental Protection in the Caspian Sea: Policy constraints and Perception. http://www.Caspianstudies.com/articles.

(5) Phillip Kurata, Caspian Ecosystem Menaced by pollution. http://ens.lycos.com/ens/apr99/1999L-04-14-01.html

(6) http://www.millennium-debate.org/ind21jun.htm

(7) Ibid

(8) http://ruzgar.aznet.org/ruzgar/1-1.htm

(9) Report of the Energy Information Administration, http://www.eia.doe.gov/emeu/cabs/caspenv.html

(10)

(11)

Iran and Nabucco Pipeline

The Nabucco pipeline is a planned natural gas pipeline from Caspian and the Middle East region to Turkey and finally to Austria. The aim of this pipeline is diversifying the current natural gas suppliers and delivery routes for Europe. The project is backed by several European states and the United States.

The main source of Nabucco's supply will be the second stage of the Shah Deniz gas field in Azerbaijan, coming on-stream in 2013. Turkmenistan would provide for Nabucco 10 bcm of gas annually. The natural gas could be transported through Iran. In the long term, Kazakhstan may become a supplier providing natural gas from the Northern Caspian reserves through planned Trans-Caspian Gas Pipeline. Egypt could provide 3–5 bcm of natural gas through the Arab Gas Pipeline. Also Iraqi gas would be imported via the Arab Gas Pipeline from the Ekas field. Iran has also proposed to supply gas to Nabucco pipeline and this was backed by Turkey; however, due the political conditions this is rejected by the EU and the United States. (Condensed from: http://en.wikipedia.org/wiki/Nabucco_pipeline)

The agreement for construction of Nabucco gas pipeline from the Caspian Sea and the Middle East to Western Europe was concluded in July 2009 in Turkey. It will be 3300 Kilometers long and it will replace part of the Russian supplies to the EU countries. It is supposed to be ready in 2014. However, it is not very clear which countries will provide the needed gas for the Nabucco.

Iran's participation is questionable because of its ongoing conflict with Washington. United States Special Envoy Richard Morningstar has said:

"I don't think there would be an agreement at this point among the Nabucco consortium for Iranian participation at this time... Our European allies, I think, are in sync with this position...This would be the absolute worst time to encourage Iran to participate in a project in Nabucco when we've received absolutely nothing in return".

Meanwhile, Nabucco Managing Director Reinhard Mitschek appears to be leaving the door open to Iran's participation. Here's what he has recently said:

"Nabucco has never excluded any source. Nabucco is not excluding any source. Bottom line, we have to buy the gas. The national gas companies will evaluate the political aspect, the commercial aspect, the technical aspect and then they will decide to buy gas from Azerbaijan, Turkmenistan, Iraq, Iran and Russia. For all these sources, we are open to transport the gas".

It will probably be years before we know whether Nabucco will buy Russian or Iranian gas, but all of this highlights the long-term energy competition looming between Russia and Iran and its implications for America's effort to secure Russian cooperation on the Iranian nuclear issue. (http://www.thewashingtonnote.com/archives/2009/07/nabucco_highlig/)

The EU's long-delayed Nabucco pipeline has received an important boost with the signing of an inter-governmental transit agreement between Turkey, Bulgaria, Romania, Hungary and Austria. With Russia's rival South Stream project having already secured the support of Italy, Serbia, Bulgaria and Greece, the Balkans is gradually becoming a tale of two pipelines. The outcome of these respective projects, therefore, will have far-reaching implications not only for Europe's long-term energy security, but for the strategic balance of the Balkans and the pressures facing the EU's enlargement agenda. (http://www.guardian.co.uk/commentis-free/2009/jul/22/gas-energy-europe-serbia/print)

In 2006, Gazprom of Russia proposed an alternative project, in competition with the Nabucco Pipeline, that would involve constructing a second section of the Blue Stream pipeline beneath the Black Sea to Turkey, and extending this up through Bulgaria, Serbia and Croatia to western Hungary. In 2007, the South Stream project through Bulgaria, Serbia and Hungary to Austria, or alternatively through Slovenia to Italy, was proposed. It is seen as a rival to the Nabucco pipeline. Ukraine has proposed the White Stream pipeline, connecting Georgia to the Ukrainian gas transport network. (http://en.wikipedia.org/wiki/Nabucco_pipeline)

Iran, while holding the second largest natural gas reserves in the world, is not a major exporter of the commodity. The EU seeks a lowering of its dependence on Russian energy, and Iran potentially could benefit by joining projects like the Nabucco gas pipeline. Iran's isolation and its poor relations with the international community are impediments that stand in the way. Iran's most important single source of natural gas is the South Pars field in the Persian Gulf, which it holds in common with Qatar. The fact that a tiny emirate across the Persian Gulf has been exploiting the gas from the Qatari side of the South Pars to the tune of billions of dollars, while Iranians helplessly witness the depletion of the reserves has caused the Iranian government a major embarrassment in the eyes of the people (Qatar enjoys the highest per capita income in the region).

Iranian politicians have claimed many times that Iran's international isolation and the economic sanctions—including those imposed by the UN Security Council—have not hurt the country seriously, and they insist on continuing the nuclear program at all costs. In reality, however, Iran's oil and gas industry (the country's main source of income) have suffered and will suffer further if no compromise is made.

The projected construction of oil and gas pipelines over the next 25 to 50 years all bypass Iranian territory and Iran will lose the transit fees, jobs, investment and prestige that accompany such projects.

The United States supports Nabucco as a means of avoiding Russian monopoly in the European gas-supply chain, and has backed the participation of Azerbaijan, Kazakhstan, and especially Turkmenistan in the project.

Brussels and Washington are also supporting the construction of a Trans-Caspian, natural-gas pipeline to run from either Kazakhstan, or more likely from Turkmenistan, along the seabed to Azerbaijan, where the gas would be pumped into pipelines leading to Nabucco. But the Russians and Iran are opposing it under the pretext of the protection of the Caspian Sea environment.

Although, in June 3 2008, U.S. Deputy Assistant Secretary of State Matthew Bryza told RFE/RL's Azerbaijani Service that while the United States backs a larger effort, including the possible export of Iraqi gas through Nabucco, it does not back the inclusion of Iran in the pipeline plans, it seems that the US and the EU are ready to accept Iran in the Nubucco, if Iran enters into some kind of compromise with the West on its nuclear question. This may be a major source to pressure the president of Iran, who has been criticized during the recent elections for his economic policies.

Conclusions:

1- Nabucco has become close to reality by the signing of the agreement in Turkey in July 2009.
2- The issue of suppliers is open to discussion yet and in fact until the pipeline is ready (it will take a few years), the issue may face a different situation.
3- Russian participation is not welcomed because it beats the purpose of construction of Nabucco(which is mainly to avoid the monopoly of the Russians).

4- Iran is able and willing to participate. The EU and Turkey hope it becomes possible. However the participation of Iran has more political side than economic side (this can be compared to the Baku Jayhan Oil Pipeline which avoided Iranian territory due to political reasons). Iran has not yet lost the opportunity, but if it does not come to a kind of compromise on its nuclear program with the West, it will definitely lose the opportunity.

Nuclear Pollution in the Caspian Sea: Introduction of several studies

Nuclear pollution is one of the less known and rarely discussed dimensions of the serious pollution problems in the Caspian Sea. At the same time, the radioactive contamination is one of the most damaging and dangerous types of pollution in the world. The nuclear activities of the coastal states, implications of the former nuclear explosions, the remnants of the nuclear tests, the nuclear wastes (which will be radioactive for thousands of years) and finally the nuclear side of oil exploration and exploitation and transportation (especially by pipelines) are the sources of nuclear danger in the Caspian Sea.

The nuclear pollution is not in the same level all over the Caspian Sea. It is different from the viewpoints of sources and types of dangers. But if we take into consideration that the Caspian Sea, as the greatest lake in the world, is not connected through natural channels to the high seas, and at the same time, a strange wind current (there are other strange characteristics in the Caspian Sea like its rising and falling levels that is still subject to controversy) keeps steering the water inside the lake like a giant spoon, then the general dangers of the nuclear pollution for all coastal states, even those who have smallest role in the nuclear contamination of the Caspian Sea (like Iran) become more evident. Here I have tried to present the interesting parts of several studies in the field of nuclear pollution in the Caspian Sea and its surrounding states. Later I will refer to the international legal instruments that can be a base for the arrangements to combat this kink of pollution in the Caspian Sea, even without determination of the legal regime of the Caspian Sea as a whole.

Discussing the effects of the nuclear pollution and exposure to dangerous levels of radiation of radioactive materials on the human beings and societies is out of the scope of present writing. However, it is in order to mention a few words here: A group of scientists headed by Professor Farrokh Djahanguiri, from Colorado School of Mines, have made a good research (http://greenaz.aznet.org/greenaz/issues98/bt7last.htm) about "Radioactive Pollution on the Apsheron Peninsula and Caspian Sea," which I am going to refer to its results later, but here, I would like to do the same thing that they have done: i.e. quoting the effects of researches related to exposure to radioactive materials conducted in the USA:

"Effects of exposure of the human body to radium is studied by health research organizations in the United States and radium is labeled a "bone seeker" because of the bone's affinity and ability to retain nuclides of that element. Bone disorders resulting from the intake of radium were documented early in the century when doctors discovered that women hired to paint luminous watch and clock dials had an unusually high incidence of bone sarcomas - a condition called osteonecrosis or "bone death".

The sensitivity of the lungs to the energy emitted by inhaled radionuclides has been documented in studies showing high incidences of lung cancer among miners of uranium-bearing ores, who became exposed to airborne particulates and concentrations of radon-gas. Most recent studies have linked increasing incidences of lung cancer to inhalation of radon, which accumulates in homes built over radon-emitting soils. With regard to the exposure pathway through which radioactive materials may be ingested, one study estimated the risk associated with a lifetime intake of drinking water containing one Pico curie per liter of radium-226 to be ten excess fatalities per 1 million members of the population. Another study relates increased incidences of leukemia to certain counties in Florida where drinking water contains elevated levels of radium."

What are the sources of nuclear pollution in the Caspian Sea?

It is surprising to say that one of the sources of nuclear pollution in the Caspian Sea is the Caspian Sea per se. Boris Gulubov (http://www.isar.org/isar/achive/gt/gt10golubov.html) reports:

"In addition to man-made sources of radiation, the Caspian ecosystem collects and stores high levels of natural radioactive nuclides. Caspian waters, bottom sediments, and living organisms contain levels of uranium five to seven times higher than those in other seas, due mostly to the complex migration patterns of naturally occurring radioactive nuclides. Mollusk fossils from millions of years ago have been found to contain much more uranium than those of their modern descendants. Because the Caspian Basin does not drain into other bodies of water, it operates as a natural precipitation tank for a significant mass of naturally occurring radioactive elements, which, once they reach the sea, have no outlet. Some of these radioactive nuclides originated in aboveground rock formations such as the granite in the nearby Caucasus, Elburz and Ural mountain ranges that are gradually eroding. Another source of nuclides includes undersea sedimentary rocks - the same strata that contain oil and gas."

But the main source of nuclear contamination danger is not nature. The nature has its own defensive systems as Gulubov has explained in his article. The past and present activities of coastal countries are the main sources of concern. R.B. Flay in his article "Trans-boundary Environmental Issues in the Former Soviet Union", dated 1998 has written:

"For reasons known to all of us, the Soviet Union (SU) developed a large nuclear industry for both military and energy purposes. While prior to 1991 this wide network of nuclear facilities was regulated by Moscow, the fact that many of the facilities lie outside of Russia is a major problem today given both the environmental and the military threat of these materials. In Russia alone there are 320 cities and 1548 other locations used to store radioactive material. In Ukraine approximately 100,000 small nuclear facilities exist and there are 11,000 in Moldova. Geologists in Kazakhstan have found about 80 million tons of radioactive waste and since the mid 1960's the Atyrau oblast has been the test site for some 17 nuclear tests. The incident of reactor four at Chernobyl in 1986 in Ukraine graphically displays the severity of this issue. Ukrainian government officials put the direct death toll of the accident at 8,000 with another 12,000 individuals being badly irradiated. Other sources place the expected death toll from cancer as a result of the accident at up to 100,000. In Ukraine, nearly 17 million acres of land was contaminated by the cesium 137 fallout, from the reactor and similar exposures were incurred by Belarus and Norway."

The role of Russians in the nuclear pollution of the Caspian Sea is substantial. I have only singled out a few cases as an example for this report. There is a great body of studies on the various aspects of nuclear pollution caused by the Russians. According to "Albion Monitor *March 30, 1996* (http://www.monitor.net/monitor):

"For more than three decades, the Soviet Union and now Russia secretly pumped billions of gallons of atomic waste directly into the earth and, according to Russian scientists, the practice continues today.

The scientists said that Moscow had injected about half of all the nuclear waste it ever produced into the ground at three widely dispersed sites, all thoroughly wet and all near major rivers. The three sites are at Dimitrovgrad near the Volga River, Tomsk near the Ob River, and Krasnoyarsk on the Yenisei River. The Volga flows into the Caspian Sea and the Ob and Yenisei flow into the Arctic Ocean.

The injections violate the accepted rules of nuclear waste disposal, which require it to be isolated in impermeable containers for thousands of years. The Russian scientists claim the practice is safe because the wastes have been injected under layers of shale and clay, which in theory cut them off from the Earth's surface.

But the wastes at one site already have leaked beyond the expected range and "spread a great distance," the Russians said. They did not say whether the distance was meters or kilometers or whether the poisons had reached the surface.

The amount of radioactivity injected by the Russians is up to three billion curies. By comparison, the accident at the Chernobyl nuclear power plant released about 50 million curies of radiation, mostly in short-lived isotopes that decayed in a few months. The accident at Three Mile Island discharged about 50 curies. The injected wastes include cesium-137, with a half life of 30 years, and strontium-90, with a half life of 28 years and a bad reputation because it binds readily with human bones...

At worst, it might leak to the surface and produce regional calamities in Russia and areas downstream along the rivers. If the radioactivity spreads through the world's oceans, experts say, it might prompt a global rise in birth defects and cancer death."

What about Azerbaijan?

Azerbaijan and certain parts of Kazakhstan, due to the previous activities (military and civilian) of the former Soviet Union and also because of the oil exploration, and exploitation activities are serious sources of nuclear pollution. The acute part of pollution by nuclear substances is regarding the nuclear wastes that large quantities of them are kept under sub-standard conditions.

N. Majidoya, in parts of his article titled: "There's no room...", Zerkalo [in Azeri], dated 25 of Mar. 2000, reports:

"The main radioactive waste storage facility in Azerbaijan is the Izotop Industrial Complex, located 30km from Baku. Izotop was constructed in the 1950s and holds 510 tanks of radioactive waste in 10 storage tanks designed to hold only low-level radioactive waste. However, as of March 2000, nine of the 10 tanks were full and, in many cases, the level of radiation is above 1,000 roentgens. Data collected before 1988 suggest that approximately 350 organizations have a total of 950 radiation sources in their possession. These organizations include military facilities, research institutes, production plants and health services-related enterprises. Spent radioactive sources from these organizations never reached the Izotop storage facility. Instead they are spread throughout the Baku region. Out of 157 radioactive contamination sites discovered in 1988 as a result of a special inspection in the Baku region, only 31 had been cleaned as of early 1996. Many of these radioactive sources were left behind by the chemical weapons divisions stationed in Baku, Lenkoran, Gyandzha and Nakhichevan during the Soviet period. Although Azerbaijan does not have any nuclear reactors, research facilities, or uranium mines, the Baku newspaper *Zerkalo* reports that the level of radiation emissions in the country is much higher than normal. The article attributes this contamination to "orphaned" radioactive sources that were left behind by the Soviet military. These include "sources from gamma ray detectors, radioactive devices, radiopharmaceutical preparations and applications, and gamma and neutron sources used in geological research." According to the article, testing has discovered 157 contaminated areas in Baku, of which only 31 have been cleaned up. The remaining contaminated areas still have radiation levels of 120-3000 microroentgens per hour, compared with the normal background level, which should not exceed 50 microroentgens per hour. Over 200 ministries, enterprises, and other institutions in Azerbaijan use radiation sources in their work, and *Zerkalo* argues that these sources are not adequately monitored, nor are necessary safety precautions being taken. The paper charges that problems are especially severe in the oil industry in Azerbaijan, saying that radiation levels of 8000-1200 microroentgens per hour have been measured at the Surakhanyneft oil and gas extraction enterprise. Furthermore, the Azeri Medical University conducted tests on oil workers and discovered that the level of radioactive isotopes in their tooth enamel was equivalent to that of residents of the Chornobyl area in Ukraine. Currently all radioactive waste in Azerbaijan, including "orphaned" sources, should be disposed of at the <u>Izotop</u> industrial complex, located 30km from Baku. The facility is almost filled to capacity, however; nine of the existing ten waste storage tanks are full, while the tenth tank is half full. The article cites the director of the facility, Baba Huseynov, as saying that "every compartment has a strictly limited capacity- 200 Curie. Today we cannot bury the ownerless sources of radiation we have found because we will thus fill up the last tank."

The authorities of the Republic of Azerbaijan are well aware of the extent and dangers of the Caspian pollution by the nuclear wastes. State Committee of the Azerbaijan Republic on Nature Protection (1993; TACIS/MoE of Georgia, 1998; Courier, Rustavi-2 night TV show, 04.02.02) has reported:

"The issue of radioactive wastes in the Caucasus region is basically related to the nuclear power plant operated by Armenia, military camps, and oil drilling and processing operations in Azerbaijan and some parts of the North Caucasus. Different research and medical institutions are also the sources for radioactive wastes. There are practically no data on these types of wastes and the issue needs to be further studied. Even if there is some information, it is frequently classified and not available for different users. Public awareness about radioactive wastes is also very low within the entire region. Therefore, casualties in the population are not rare. In particular, a high threat is from former Soviet military bases, where significant amounts of radioactive wastes are accumulated. There are no comprehensive inventories of radioactive sources and wastes. Nor do storage facilities exist for them. Although, the Caucasus countries have designated authorities, they have little capacity to handle the issues. In Azerbaijan, radionucleides of naturally occurring radium, thorium and potassium were found in oil drill fields. At some places, soils are so polluted that they need to be buried as radioactive wastes. "oil lakes" and flood fields, created while pumping bore-waters back into oil-bearing layers, aggravate the situation. Some old oil drill fields currently are used as settlements hence the population is exposed to radon noble gas damaging to the lungs. A similar situation exists for chemical plants and oil refineries. Ground waters with high radium-226, thorium-228 and potassium-40 content were used in a Baku iodine plant as a raw material. Consequently, part of plant territory and equipment were polluted by radionucleides. Especially urgent is the problem of activated charcoal decontamination, accumulated in the plant territory."

The case of nuclear pollution through the Azerbaijan Republic, especially in the Apsheron peninsula, (it is very close to Iran and it is the same place that Azeri authorities have proposed to Americans, according to several reports, for creation of a military base) has been the subject of a valuable study headed by Professor Farroh Djahanguir (previously refereed to this report and its address) under the title of "Radioactive Pollution on the Apsheron Peninsula and Caspian Sea." Some of its contents reveal that:

"Radioactivity in the oil fields of Azerbaijan was reported by the Geology Institute (GI) of the Azerbaijan Academy of Sciences and the State Committee for the Nature Protection (NP). However, the environmental impact of radioactivity on soil contamination, surface and ground waters, and aquatic life is extensive. Oil production from the Caspian Sea's on-shore and offshore oil fields dates back to the late 1800's.

Oil production from oil fields of the Apsheron Peninsula started at the beginning of the 19th century. During the intervening years, radioactive contamination of soil, water, and air have occurred. In the recent official Azerbaijan publication on nature protection, it is stated "research results of study of radionuclide composition of tests have showed elevated presence of natural radium, thorium and potassium". It has been determined again by Azerbaijan officials during all phases of oil related activities such as production, processing, transportation, and storage, that radioactive contamination of the environment has occurred. Additionally, it has been stated that injection of the produced water into oil-bearing strata has caused severe radioactive contamination of the environment. Additional sources of contamination are reported from handling and service of the oil field equipment, chemical processing and oil-refinery plants which are located near Baku and Sumgait . For example, the Baku iodine plant, during iodine production, elevated high concentration of radionuclides (namely, radium-226, thorium-228 and potassium-40) and discharged them into water.

According to compiled data reported by NP, total volume of contaminated water discharged to surface, ground water, and the Caspian Sea in 1991 was 2.9x108 m3/year. This data comparison to 1985 shows a four-times increase.

This contaminated water is a major source of radioactive pollution in the environment.

It should also be stated that some of these contaminates sites are used for settlement of people and that most recently Azerbaijan refugees, displaced from their own homes by the Armenian occupation, have been relocated to these contaminated sites.

It has been stated that environmental contamination in the Republic of Azerbaijan, due to the presence of NORM in oil fields, oil equipment and processing plants are close to a national disaster.

Study of radionuclide composition of soils, water, and hydrocarbons in Apsheron Peninsula (AP) shows presence of elevated radium, thorium, and potassium. Radioactivity background varies in different locations in the world. It ranges from 3-60 mR/hr. In some locations at AP, radiation from natural ionizing sources exceeds radioactivity background 10 to 100 times. Geological formations at AP are composed of clays, sandstone, and limestone with radioactivity background within 6 mR/hr. At the Caspian Sea coast, sea-sand's radioactivity background decreases to 3 mR/hr. At the tectonically disturbed zones of AP, radioactivity background increases to 15-20 mr/hr. The average radioactivity background in Azerbaijan according to GI radiation experts is within 10-12 mR/hr. At some oil fields in South-western AP and around iodine plants the level of radioactivity elevated background varies from 600 to 5.000 mR/hr. The origin of NORM at oil and gas fields of AP is related to drilling and production of oil and gas.

Other sources of radioactive contamination of the environment in Azerbaijan are mud volcanoes and geothermal energy production wells. Radioactive contamination of the environment is not limited to Azerbaijan; other neighboring countries namely, Russia, Kazakhstan, Turkmenistan, and Iran are facing radiation related environmental problems."

Perhaps the most important study is the work of group pf scientists headed by B. Shaw (http://www.grida.no/caspian/additional info/environment baseline 1pdf). This study must be the center of every research about the nuclear pollution in the Caspian Sea. Let's look at some of its contents (some of the figures and tables that are mentioned in the following section are not included in this report. However, they are readily available in the address of the report):

"In the Caspian Sea region of Central Asia, there are several nuclear reactors used for power production and research, and many nuclear sites remaining from activity of the FSU [former Soviet Union], including those of uranium mining and production, nuclear waste dumping, storage, fuel production, and PNEs. The Caspian basin includes all of the sites north and beyond Moscow to the headwaters of the Volga River, the to the west to the source of each of the major river systems, the Kama, Ural and Emba, and the Kura to the east of the sea. However, for the purposes of this study, only those within the immediate Caspian Sea region are described in detail because of their direct potential impact to the sea and to the region's vulnerability to transnational conflict (B. Shaw, personal communication, June 1998; D.J. Bradley, personal communication, June 1998).

Figure 19 and Table 4 indicate the sites of nuclear reactors for the production of power in the vicinity of the Caspian Sea, sited at Novovoronezh (Volgadonsk) (Figure 20), Balakovo (Figure 21), and Rostov in Russia, at Aqtau in Kazakhstan (Figure 22), and at Yerevan in Armenia (PNL 1998a, 1998b; INSC 1997, 1998b, 1998c, 1998g). Reactors for research are found at Tehran and Esfahan, Iran, and Dmitrovgrad, Russia (EIA 1998; INSC 1998a, 1998c). Others outside the immediate Caspian Sea area that could potentially have an impact to the sea, for example, via the Volga watershed, are probably at low risk of doing so. Nonetheless, a recent report from the Bellona Institute (Kudrik 1997) reported accidental discharge of radioactivity to the atmosphere at the Dimitrovgrad Research Institute on the Volga River July 25-26, 1997. Discharge levels of [131]iodine were about 18 times above the normal levels (2.2 to 2.6 GBq for two days, 1.9 to 2.2 GBq for 5 days, above the regular levels of 122 MBq/day) for about one week (Kudrik 1997).

All phases of the nuclear fuel cycle, along with weapons testing, accidents, deliberate discharge of wastes, and disposal of industrial, medical, and research wastes could potentially contribute to radionuclide contamination of the Caspian Sea. Nuclear fuel cycle activities include past and present uranium mining and milling operations, uranium conversion, enrichment and fuel fabrication, irradiation in nuclear reactors, and storage of wastes from every step in the cycle.

Table 4. Nuclear Reactors in the Caspian Sea Vicinity

Country	Location	Reactor	Purpose	Reference
Armenia	Madzamor (Yerevan)	PWR[a]	Electricity	INSC 1997
Iran	Tehran	Unknown	Research	EIA 1998; INSC 1998a
	Esfahan	ENTC GSCR[b] ENTC HWZPR[c] ENTC LWSCR[d] ENTC TRR[e]	Research	INSC 1998a
Kazakhstan	Aqtau	LMFBR[f] (BN-350)	Electricity	INSC 1998b
Russia	Novovoronezh (Volgadonsk)	PWR (7 units)	Electricity	PNL 1998a
		VVER[h] (210 through 1000)		INSC 1998e

	Rostov (Volga-donsk)	PWR VVER (4 units)	Electricity	INSC 1998g
	Balakovo	PWR (4 units)	Electricity	PNL 1998b
		VVER-1000		INSC 1998f
	Dmitrovgrad	BWR[i] (4 units)	Research	INSC 1998c

a) PWR pressurized water reactor.

b) ENTC GSCR subcritical water reactor.

c) HWZPR tank-in-pool heavy water reactor.

d) LWSCR subcritical light water reactor.

e) TRR pool water reactor.

f) LMFBR liquid metal cooled fast breeder reactor.

g) EWG-1 tank-type water- and gas-cooled reactor.

h) VVER is a Soviet-designed PWR, in Russian called Vodo-Vodyanoi Energeticheskii Reaktor.

i) BWR boiling water reactor; at Dmitrovgrad, there are four different research BWRs: MIR-M1 (channels and pool); RBT-10/1 (pool); RBT-10/2 RBT-6 (pool); SM-2 (tank).

The major problems related to waste management that are reported for Russia, but which would likely apply to the other republics of the FSU as well, are as follows: large quantities of existing and newly generated radioactive wastes remain untreated; a lack of facilities for safe handling of radioactive waste and spent nuclear fuel; facilities that are not considered safe, do not meet current environmental requirements, and/or are filled to capacity. These problems increase the risk of radioactive contamination of the environment and for radiation accidents.

Although sites of nuclear activity are not as concentrated in the Caspian Sea region as in some other areas of the FSU, there are nonetheless a number of sites of potential concern. Near the Caspian Sea, Armenia, Azerbaijan, Russia, and Turkmenistan all contain regional radioactive material storage sites, called *radons*. For example, the radon at Baku, Azerbaijan, on the shore of the Caspian, does not treat wastes, but stores up to 25 m^3/year of solid and liquid radioactive waste materials. The radon facility at Yerevan, Armenia, stores up to 5 m^3/year of solid, liquid, and biological radioactive wastes, and spent ionizing radiation sources. Rivers of northeast Azerbaijan flow directly into the mid-Caspian Sea; rivers of southeast Azerbaijan, the major one of which originates in Armenia and drains the Yerevan area, flow directly to the south Caspian. Therefore, any radioactive wastes carried from mining or former-processing sites in these areas would also potentially be carried to the sea (ENRIN 1997b). Further, former uranium mining and processing sites and sites of PNEs are in the region of the Caspian and could pose risk for release of radioactive materials into the waterways that lead to the sea (Figure 23, Table 5). A detailed description and inventory of radioactive residues and wastes resulting from the FSU nuclear activities in this region can be found in Bradley (1997).

On the Turkmenistan coast of the Caspian, two chemical factories that use activated charcoal in their industrial processes have released radioactive wastes onsite at Cheleken Chemical Factory and Nebit Dag Iodine-Bromide Factory (Figure 23, Table 5). The total radioactive pollution at the former site has been monitored at 200,000 Bq/kg (average 80,000 Bq/kg) of wastes, in a total of 15,000 to 18,000 mt of wastes that are accumulated around the factory (Berkeliev 1997), which would equal a total maximum activity of about 40 Ci (D. Bradley, personal communication). There are also deposits of radiobarites in old wells drilled for oil, gas, and industrial salts at Cheleken, the total radioactivity of which was estimated at 10 million Bq (.0003 Ci) in 1966 (Berkeliev 1997).

Although detailed information is not available, it is strongly suspected that PNEs were carried out for industrial purposes at least once in 1972 in the Mary Region of Turkmenistan to seal a gushing petroleum well, and similar PNEs were carried out in the Ustjurt and the Kyzlkum of Kazakhstan near the Turkmenistan border (Berkeliev 1997; Bradley 1997; Figure 23, Table 5).

Table 5. Nuclear Fuel Processing Facilities, Radons,[a] and Other Potential Sources of Radioactive Pollution in the Caspian Sea

Country	Location	Facility	Facility type	Reference
Armenia	Yerevan	Radon[a]	Radioactive waste storage	Bradley 1997
Azerbaijan	Baku	Radon	Radioactive waste storage site	IAEA 1995 (in Bradley 1997)
Kazakhstan	Mangyshlak	Uranium strip mine	Uranium processing	Berkeliev 1997
	Mangyshlak	Underground nuclear test site	Peaceful nuclear explosions (three)	Bradley 1997
	Aqtau	Kaskor uranium mill	Uranium tailings	Bradley 1997
	Plato Ustijurt	Underground nuclear blast site	Peaceful nuclear explosion	Bradley 1997
	Sarykamys area	Underground nuclear blast site	Peaceful nuclear explosion	Bradley 1997
	North shore Caspian near Kazakhstan western border	Underground nuclear blast site	Peaceful nuclear explosions (series)	Bradley 1997

Russia	Novovoronezh, Volgadonsk	Novovoronezh Reactor site	Spent fuel storage	INSC 1998d
	Lermontov	Uranium mine	Uranium mine	Bradley 1997
	Volgograd and Samara on the Volga River Dmitrovgrad	Radon Dmitrovgrad Research Institute	Radioactive waste storage site Radioactive waste injection	Bradley 1997 Bradley 1997
	Dmitrovgrad	Dmitrovgrad Research Institute	Accidental discharge	Kudrik 1997
Turkmenistan	Cheleken	Cheleken Chemical Factory	Industry using activated charcoal	Berkeliev 1997
	Nebit Dag	Nebit Dag Iodine-Bromide Factory	Industry using activated charcoal	Berkeliev 1997
	Kizilkaya	Gyusha transfer station	Uranium mining, transfer	Berkeliev 1997
	Karakumskij Canal, Ashkabad		Radioactive waste storage site	Bradley 1997
Uzbekistan	Kyzlkum near Kazakhstan border	Underground nuclear blast site	Peaceful nuclear explosion	Berkeliev 1997

a) Radon is a regional radioactive waste storage site in the FSU republics.

Situation of Kazakhstan is horrible.

Cynthia Werner from Department of Anthropology of the Texas A&M University reports:

"Between 1949 and 1989, the Soviet government conducted more than 470 nuclear tests at the Semipalatinsk Nuclear Test Site (SNTS) in northeastern Kazakhstan. The test site is surrounded by several villages and located approximately 150 km west of the city of Semei (formerly Semipalatinsk). According to some estimates, up to two million people living in the region have been exposed to varying doses of radiation as a result of these tests."

Also, the report of National Nuclear Center of Kazakhstan (http://www.tech-db.ru/istc/db/projects.nsf/prjn/k-632) shows:

"The ecological situation in the Mangystau province (Kazakhstan) became aggravated in 1960's, at start of exploration of the deposits of uranium ores, oils, and raw minerals and the creation of a chemical industry in Mangyshlak, in neglect of environmental problems. The KOSHKAR-ATA tailing pond is the most hazardous place among all objects, making a considerable contribution to atmospheric contamination with powder radioactive and toxic wastes of chemical and mining metallurgic industries. KOSHKAR-ATA represents a serious hazard for habitants of Aktau and adjacent inhabited localities.

The KOSHKAR-ATA tailing pond, a drain-free settling pool for industrial, toxic, chemical and radioactive wastes, and for ordinary domestic drains, is 5km north of Aktau (Mangystau province), which is situated on the shore of the Caspian Sea. Industrial, toxic and radioactive wastes, solid sediments of unpurified ordinary domestic drains from a part of the Aktau residential region have been placed in the tailing pond since 1965 and have been stored there up to this day.

Solid radioactive wastes of the chemical mining metallurgic plant, where uranium ores were processed, were buried without control or official account in a trench-type burial without hydro-isolation. According to the data of the Mangystau Provincial Ecology Department (Aktau), the real mass of radioactive wastes (RAW), disposed in the tailing pond, is about 360 million tons with 11000Ci total activity. Results of works, which were carried out by the Institute of Nuclear Physics in 1999, showed that the exposure dose rate (EDR) at shallow zone equaled 80-150mkR/h. Some places were revealed, where EDR was 1500mkR/h and radionuclide content was up to 548-5000bk/kg. According to preliminary experimental data on samples from KOSHKAR-ATA, which were obtained by using a certified EPR-dosimetry method, dose load measured 15-25kGy. That is similar or even exceeds the dose value of soil samples from the Southeast trace of the Semipalatinsk test site. As a result of a steady reduction in the water phase level, the area covered by just bed sediments, which are a source of toxic dust, has recently increased. Under existing hydro-geological conditions of the tailing pond region, there is potential for penetration of liquid waste to aquifers and to the Caspian Sea."

Boris Gulubov which I refereed to his article earlier, has written in other parts of his interesting report:

"When I served on a team that conducted geological surveys in Central Kazakhstan in the early 1960s, we frequently received strict orders to temporarily stop searching for traces of uranium. However, out of curiosity we occasionally ignored the political leadership, turned on the radiation meters, and watched as the indicator arrow flew off the scale, showing radiation intensity hundreds – if not thousands – of times higher than naturally occurring levels. The cause? Clouds of radioactive dust kicked up by powerful atomic explosions over the Semipalatinsk testing ground, a few hundred kilometers away.

However, during my research on the Caspian Sea at the National Institute for Marine Geology and Geophysics in 1966, my belief in the political propaganda about nuclear explosions began to wane after several of my friends who had worked near the Caspian for many years died suddenly of leukemia. I became convinced that, rather than isolated incidents, such cases were typical along the Caspian coast. At the same time, articles began to appear in the press about the danger of radioactive fallout from nuclear explosions.

By 1967, I had learned that the Caspian's levels of bomb-grade tritium – a highly enriched radioactive element, dangerous to all living creatures – had increased 300 to 400 times since the atomic blasts. The half-life of this isotope is 12.8 years, so it was not surprising that into the 1970s high levels of tritium were found in Caspian water samples. Data published recently by both the Russian NGO Taifun and the Russian government's Hydrological and Meteorological Center confirm the high level of radioactive contamination at the time.

From 1966 to 1987, 69 underground industrial nuclear explosions were detonated in the vicinity of the Caspian Sea. Twenty-four of these were designed to create underground storage chambers in the salt domes of the Astrakhan, Karachaganak, Orenburg, and Sovkhoz condensed gas deposits. Experts also hoped nuclear explosions would increase the productivity of oil sites at the Grachevskii and Takhta-Kugultinskii fields."

It is necessary that the countries of the Caspian Sea region try to establish a system for combating the nuclear pollution in this area as soon as possible. The role of the atomic energy organizations in each of these countries and also the International Atomic Energy Agency in this field is very important. Fortunately, at the moment, there are several major international instruments that can be used by the concerned parties for protection of the Caspian Sea against the nuclear pollution. I believe even before the legal regime of the Caspian Sea is finally determined among the coastal countries, serious steps must be taken in this field. Some of the most important international instruments that must be observed in the process are:

1-   Vienna Convention on Civil liability for Nuclear Damage, 1963 and its 1997 Protocol.

2- London Convention on the Prevention of Marine Pollution by Dumping of Wastes and other Material, 1972.

3- Convention on Physical Protection of Nuclear Materials, 1980.

4- Monte go Bay (united Nations) Convention on the law of the seas, 1982.

5- Basel Convention on the Trans-boundary Movement of the Hazardous waster and their Disposal, 1989, and its Protocol of 1999 on Liability and Compensation.

6- Vienna Convention on Nuclear Safety, 1994.

7- Vienna Convention on Supplementary Compensation For nuclear damage, 1997.

8- Helsinki Convention on the Protection and Use of Trans-boundary Waters and International Lakes, 1992, and its Protocol of 1999.

9- Vienna Joint Convention on the safety of Spent fuel Management and on the Safety of Radioactive Waste Management, 1997. This Convention is especially important for transportation of nuclear materials taking into consideration the recent arrangement between Iran and the Russian Federation for sending the spent fuels of the Bushehr Nuclear Power Plant and possibly other power plants in Iran. The negotiation for construction of 4 to 10 other nuclear power stations in Iran by the Russians are under way, according to the strategic program of cooperation of Iran and Russian Federation.

I have two important points to add here:

A- Out of the several thematic centers that are established by the CEP (Caspian Environment Programme) in the littoral states of the Caspian Sea, the Legal Center, which is in charge of preparing regulations, is in Moscow. I do not think that Russians are very interested in preparing regulations which most of them would address themselves. May be these centers should circulate among the concerned states, before becoming fully independent from the CEP.

B- In preparing legal documents and operational standards in the Caspian Sea, due attention should be given to the current international regulations and standards regarding the special areas. In these areas (such as the Antarctic waters) in addition to the general rules and regulations designed to protect all environments, some particular regulations are in place because of the special geographical or physical characteristics of the areas. In the case of the Caspian Sea, the fact that this body of water is not really connected to the open seas of the world makes it imperative to have special rules and standards.

Iranian Doctrine of the Caspian Sea

Iran is planning to define its Caspian Sea Doctrine. The concerned doctrine has several components which are designed to answer the views of the Iranian government in the various fields starting from the legal regime of the Caspian Sea and extending to the environment, shipping, and flying over, exclusive economic and fishing areas, joint ventures in the economic activities and the security of the region. The key words in the "Iranian Caspian Sea Doctrine" are: "consensus", "equity" (in the divisions) and at the same time, "keeping the non-littoral states out" of the region.

Mohammad Mehdi Akhunzadeh, the Deputy Iranian Minister of Foreign affairs and also the special envoy of the Iranian president in the Caspian Sea affairs, following the recent Summit of the Caspian littoral states in Baku Azerbaijan Republic (18 November 2010), which was the third summit of these states (the first one was in Ashgabat 2001, and second one in Tehran 2007), has said that Iran is preparing its Caspian Sea Doctrine.

On the basis of what he has told as the report of the Baku Summit to the Iranian media, and also other reports about the positions of Iran regarding the Caspian Sea, it is possible to conclude that that the main points in such a doctrine are as follows and at the same time, some comments are made about each of them.

1- The Caspian Sea is a sea of "peace and friendship". These words have been repeated by many Iranian officials, including the president and foreign minister, before and after the Baku Summit. This point has in fact two important sides: it is an expression of hope that peace prevails in the Caspian Sea and at the same time, it is indirectly a kind expression of concern that the Caspian Sea has potentials for conflict (1) and if the littoral states cannot find ways to solve their problems from peaceful ways, the possibility of conflict is there. Lack of attention to the fundamental interests of a major power in the region (noting that the other states of the Caspian Sea are increasingly becoming careless about Iran's positions in the Caspian Sea) can be a cause of problem to peace and stability. This is also a call for refraining from militarization of the Caspian Sea. Although this is an important point, but probably the time to stop it has passed.

2- The case of the Caspian Sea is a "Sui generis" (a special case). This is a key issue in the Caspian Sea affairs and it affects all other issues related to this region. The Caspian Sea is a unique case and for the same reason its legal regime and the delimitations and other maritime issues are not subject to the general rules of the international law of the sea. This feature gives the littoral states of the Caspian Sea the ability and the right to find their own formula for the problems of the Caspian Sea. In the same context, some of the littoral states (such as Kazakhstan) have proposed in the past that at the same rules of the international law of the sea to be used in the Caspian Sea. However, the other states are under no obligation to accept this suggestion. The case of the Caspian sea has not been discussed in important international law of the sea occasions (such as the United Nations conferences on the law of the sea and especially the Third UN Conference on the Law of the Sea which resulted in 1982 UN Convention) because of the special and exclusive status of this body of water.

3- No legal regime can be defined in the Caspian Sea without agreement of Iran. Noting that some states of the Caspian Sea, led by the Russians, have opted for delimitation of the Caspian Sea-bed by Modified Median Lines (MML) and Iran is basically opposed to this formula (the reason is that the formula gives the smallest share to Iran, among others.) (to see the implications of the MML and the reasons of Iran to oppose it please see the article mentioned in note number 2)

4- The legal regime should be defined on the basis of "equity" as a principle of the International law. Iran's position regarding the legal regime of the Caspian Sea has gone through several periods. (3) In the first period Iran was supporting the common administration of the Caspian Sea. Later, when confronted with the rejection of this formula by others, Iran started to ask for the 20 percent for each. Yet, in another stage Iran asked for the division on the basis of "equity". During the last couple of years, Iran has been talking about the "equity", but now, especially before the Baku Summit of 18 November 2010 and after that, it seems Iran means something other than what it was at least implied to mean by this criteria. The new special envoy of the Iranian president for the Caspian Sea, immediately after the conclusion of the meeting of the Caspian Ministers in Tehran (15 November 2010), and a few days before the Third Summit of the Caspian States in Baku (18 November 2010) in response to a question by the official news agency of Iran about the 20 percent share of Iran in the Caspian sea, responded: our aim goes further than this limit. This indicates to a new change in the Iranian policy because up to that date, what was generally understood from the insistence of Iran on the principle of equity in the Caspian Sea was that it may agree to something less than 20 percent, provided that the share of Iran is extended from 13 to couple numbers higher (for example 17) and especially to include some of the places favored by Iran (such as Alborz oil fields that Azerbaijan Republic calls it Aloo). According to the new policy, Iran's interpretation from the equity is at least 20 percent of the entire Caspian Sea if not more. (4)

5- Any decision for the legal regime of the Caspian Sea must be the result of "consensus "among the littoral states. This point is based on the fact that the littoral states should decide the fate of the Caspian Sea and it is necessary that they agree on it. The issue was first mentioned in the earlier agreements of the littoral states but later mentioned in clear words in the final decoration of the Tehran Summit in 2007.

6- The non littoral or third states should not interfere in the issues of the Caspian Sea. The point is based on the old Iran-Russian treaties of 1921 and 1940. These treaties, although they are old and not accepted by some of the newly independent countries, still form the backbone of the legal regime of the Caspian Sea until such time that the littoral states find a new regime agreed by all of them for the Caspian Sea. Article 13 of the 1940 "Treaty Regarding Trade and Navigation between the USSR and Iran", provides that : " the contracting parties agree that in accordance with the principles set forth by the Treaty of February 1921, between the Russia and Persia, only vessels belonging to the USSR or Iran and likewise to nationals and commercial and transport organizations of either of the contracting parties , sailing under the flag of the USSR or Iran respectively may be found throughout the Caspian Sea." (5) this point has been emphasized, especially by Iran (mainly due to the concerns that the littoral states may put parts of the Caspian Sea at the disposal of non-littoral states to stage an operation against the present regime of Iran) in all meetings of the littoral states and it was reflected in the final dictation of the Tehran Summit in 2007.

7- Security of the Caspian Sea will be provided by the littoral states. The Caspian Littoral states have signed an agreement regarding this issue (Baku 2010) and they are going to conclude more protocols on it later. The main titles of this agreement are: combat against terrorism, combat against organized crimes, combat against smuggling of arms and narcotic drugs, combat against money laundering, human trafficking, illegal fishing and security of shipping and combat against piracy. (6)

8- Economic cooperation of the littoral states should be expanded through establishment of joint companies among the littoral states. Iran was after this idea since the start of the new era in the history of the Caspian Sea, which was the collapse of the USSR and emergence of the newly independent countries around the Caspian Sea. However, the realization of this idea is far from reality due to the different political orientations of the littoral states and at the same time the policies of Iranian regime which have made the country isolated.

9- The environment must be protected against all sources of pollution. Pollution in the Caspian Sea is a major issue. Apart from the unregulated and illegal fishing which has resulted in endangering the Caviar producing species, exploration, exploitation and transportation of the oil and gas in the Caspian Sea have put it to new dangers. The littoral stats have concluded a convention in 2003 (Tehran Convention) to combat the pollution but they have not made much progress. (7) Iran believes laying oil and gas pipelines on the seabed of the Caspian Sea can leave negative impacts in the Caspian Sea and it should be avoided to the extent possible. However, it seems this stance is directly related to the level of participation of Iran in such projects including the new gas pipeline from the Caspian Sea to the EU, called Nabucco (8). Iran has been excluded from many of such projects due to the bad relations with the Western countries and especially on the nuclear program of Iran. (9)

10- Determination of a 25 mile exclusive zone for littoral states. This idea has been discussed in many meetings of the Caspian littoral states and different figures have been proposed. According to the 1940 Treaty of Iran and Russia, a ten mile "exclusive fishing zone" was determined in the Caspian Sea for the two sides. In the Baku Summit, the idea of a 25 mile exclusive zone was supported and the concerned states are going to make decisions on it. Iranian officials have talked about a 12 mile being like a territorial sea (10). 1982 law of the Sea convention gives the right to states to claim 12 miles or less for the territorial sea, but here there is no obligation to follow that rule. Also the same officials have refereed to another 12 mile or so as the "exclusive fishing zone". The idea of exclusive fishing or economic zone in the Law of the Sea convention extends up to 200 miles if possible from geographical point of view. Therefore, it can said the idea of a 12 or 13 miles exclusive fishing zone, is a special feature for the Caspian Sea and in spite of similarity with the expressions used in the 1982 UN Convention on the Law of the sea, this has nothing to do with that.

Notes and sources:

(1) Refer to the "http://www.payvand.com/news/10/dec/1061.html "

(2) Please refer to : http://www.payvand.com/news/09/sep/1102.html

(3) Bahman Aghai Diba, Law and Politics of the Caspian sea, Booksurg, 2006

(4) http://www.payvand.com/news/10/nov/1211.html

(5) Ibid, Law and Politics of the Caspian sea, Booksurg, 2006

(6) http://wap.trend.az/fa/page1/1786162

(7) Caspian Sea is Dying, http://www.payvand.com/news/10/nov/1032.html

(8) Iran and Nabucco http://www.payvand.com/news/09/nov/1240.html

(9) http://www.payvand.com/news/10/dec/1061.html "

(10) Iranian special envoy on the Caspian Sea affairs http://wap.trend.az/fa/page1/1786162

Iran-Iraq: Shatt al-Arab or Arvand Rood Dispute

Shatt al-Arab is the name given to the 200 km long, 400-1500 m wide combined effluent of the Euphrates and Tigris below Qurna, to which the Karun (Kārūn) contributes from Khorramshahr. (1) Iranians call it Arvandrood. The river is now part of the Iran-Iraq borders. The length of border between the two countries is almost 100 miles. The river is one of the navigable rivers of the world. The depth of its water is 10 to 25 meters generally, but it is 45 meters in the deepest point. The river has full capacity for shipping. Export of oil and gas from Iraq And Iran has given special importance to it. Aroung Arvanrood, there are biggest date plantations of the world.

Unlike what some people in Iraq and even in Iran think, Arvand is not a name fabricated by the former regime of Iran in order to confront the using of a false name for the Persian Gulf" by the Arabs. In fact, the history of this river is more than the Islamic period in the region. At some juncture, the two parts that combine to form the Arvand rood, joined and before that, the two rivers used to flow separately into the Persian Gulf.

Arvand was the ancient name for Tigris in Persian. There are many pieces of poems and literary works in ancient Iran that have made a reference to Arvand. For instance, Ferdowsi, the famous epic poet of Iran, has mentioned in his well known book "Shahnameh" (Book of Kings), that :

"If you do not know the language of Pahlavi [Old Iranian language], you may call Arvand as Dejleh [Tigris] in Arabic."

In another poem he says : "he [the king] went towards Arvand-drood, because he was pursuing dominance" .

Also, referring to the dream of an Iranian king [Anousheera-van], he says : " a rage started from Arvandrood and they told the religious leader that this is good omen."

Dejleh is the Arabic version of Tigris and it starts from the Turkish mountains. Due to its mountainous course, it was called " Tiz" (quick) in Persian. The ancient Iranians had various reasons including religious ones, to respect Arvandrood. The river is one of the seven holy rivers in Avesta 9 the holy book of the Zoroastrians), and according to the ancient Iranian religious narrations, the Zoro-astrian Messiah, (Soshiyant) will emerge in the shores of this river.

In fact, Shaat al Arab is rather a new name and it has not a very long historical precedence. Hamdollah Moustofi, the author of "Nozhat-tol-Gholoob" (4) has said : " the water that came from Khuzestan joins and forms Shatt al Arab and the Persians call it Arvandrrod."

As an international River, the conditions of this river fully are compatible with the definitions of such bodies of water.

The 1921 Convention of Barcelona regarding the legal regime of the international rivers provides in article one that:

"In the application of the Statute, the following are declared to be navigable waterways of international concern :

1. All parts which are naturally navigable to and from the sea of a waterway which in its course, naturally navigable to and from the sea, separates or traverses different States, and also any part of any other waterway naturally navigable to and from the sea, which con-nects with the sea a waterway naturally navigable which separates or traverses different States.

It is understood that:

(a) Transshipment from one vessel to another is not excluded by the words "navigable to and from the sea";

(b) Any natural waterway or part of a natural waterway is termed "naturally navigable" if now used for ordinary commercial navigation, or capable by reason of its natural conditions of being so used; by "ordinary commercial navigation" is to be understood navigation which, in view of the economic condition of the riparian countries, is commercial and normally practicable;

(c) Tributaries are to be considered as separate waterways;

(d) Lateral canals constructed in order to remedy the defects of a waterway included in the above definition are assimilated thereto;

(e) The different States separated or traversed by a navigable waterway of international concern, including its tributaries of international concern, are deemed to be "riparian States."

2. Waterways, or parts of waterways, whether natural or artificial, expressly declared to be placed under the regime of the General Convention regarding navigable waterways of international concern either in unilateral Acts of the States under whose sovereignty or authority these waterways or parts of waterways are situated, or in agreements made with the consent, in particular, of such States. (5)

Also, it fair to say that the modern international law on the common bodies of waters between the states leans towards equity.

"It is fair to say that the law of international watercourses has developed around the doctrine of equitable utilization, authoritatively formulated as a substantive legal principle of water allocation by the International Law Association in its 1966 Helsinki Rules on the Uses of International Rivers.41 More recently, article 5 of the 1997 UN Convention on the Law of the Non-Navigational Uses of International Watercourses has restated it as the overarching principle governing the utilization of international watercourses. In 1997, the customary status of the sovereign right to utilize an international watercourse in an equitable and reasonable manner as the fundamental substantive legal norm was also confirmed by the International Court of Justice in the case concerning the Gab-cikovo-Nagymoros Project between Hungary and Slovakia." (6)

The dispute of Iran with the Ottoman Empire, the British who had the present day Iraq's control under the mandate system (which was established after collapse of the Ottoman empire and its disintegration), and eventually the government of Iraq has a long history. At the same time, this border river dispute has attracted many studies and it can be claimed to be one of the most widely covered cases of the international river disputes in the history of the modern international law.

In 1960, E. Lauterpacht, one of the best known experts of the international law in the world, who words has turned into a source of the international law, wrote an article about the Shatt al-Arab or Arvand Rood Dispute. The title of his article was "Legal Aspects of the Shatt-a-Arab frontier." (2) he was one of the first persons to speak about the definition of Thalweg.

In 1975 Iran and Iraq concluded 1975 treaty and accepted the Thalweg as the border of Iran and Iraq in the river.

In 1981, following the invasion of Iran by Iraq, and the Iraqi claims about abrogation of the 1975 Treaty (Saddam throw a copy of the treaty into fireplace before attacking Iran and after capturing some parts of Iran, he set several conditions for the peace and one of them was abrogation of the 1975 treaty.). Kalid Al Aziz, an Iraqi lawyer claimed in a study that because Iran's actions, the 1075 treaty has lost its credibility. (3)

Notes:

(1) http://www.iranica.com/articles/shatt-al-arab
(2) http://www.jstor.org/pss/4283303
(3) ibid.
(4) vol.3, pp 214-215
(5) http://www.legislation.gov.hk/doc/multi_904v1.pdf
(6) http://www.sam.gov.tr/perceptions/Volume8/March-May2003/MeteErdem.pdf

Attach:

What is the Algiers Accord of 1975?
The Algiers Accord of 6 March 1975 was concluded between the Shah of Iran and Saddam Husien of Iraq and it was in fact the father of 1975 Treaty of Iran and Iraq. Calling the 1975 Treaty as the "Algiers Accord" is only a common (journalistic) mistake. The 1975 Treaty of Iran and Iraq was actually signed in Baghdad in 13 March 1975.

In March 6 1975, upon the initiative of President Houari Boumedienne, the Shah of Iran and Saddam Hussein (at that time, the Vice-Chairman of the Revolution Command Council, but the strong man of Iraq) met and conducted lengthy talks on the relations between Iraq and Iran. They decided: First: Carry out a final delineation of their land boundaries in accordance with the Constantinople Protocol of 1913 and the Proceedings of the Border Delimitation Commission of 1914.

Second: Demarcate their river boundaries according to the Thalweg line.

Third: restore security and mutual confidence along their joint borders. They provided: they shall also commit themselves to carry out a strict and effective observation of their joint borders so as to put an end to all infiltrations of a subversive nature wherever they may come from.

Fourth: The two parties have also agreed to consider the aforesaid arrangements as inseparable elements of a comprehensive solution. Consequently, any infringement of one of its components shall naturally contradict the spirit of the Algiers Accord. (4)

What is the 1975 Treaty?

The Foreign Ministers of Iran and Iraq met in the presence of Algeria's Foreign Minister in March 1975 and set up arrangements for the Iraqi-Iranian joint commission.

"In a practical sense, granting Iran half the river in return for an end to a crippling civil war was a cheap price for Baghdad to pay; yet Iraqi leaders felt profoundly humiliated by this accord, which ran so directly counter to the Ba'th party's intense nationalist ideology. Having to give up this part of the "Arab nation" to Iranians was not much better in their eyes than conceding Palestine to the Israelis." (5)

Article 4 of the 1975 treaty provides: the two sides agree that the protocol as and annexes of the treaty are the inseparable parts of it. These provisions are final, permanent and inviolable and inseparable elements of a comprehensive solution. Therefore, violation of any part of the treaty is contrary to the "Algiers Accord." (6)

Why Talebani hates this treaty?

In the early 1970s, the Kurdish rebellion against the central Iraqi government (Saddam) had reached a critical point. The Kurds, supported by Iran (and the USA) had put Saddam's regime in the verge of collapse. In fact, Saddam was forced to drink the chalice of poison when he accepted the Algiers Accord. The accord saved his regime, Iran stopped supporting the Kurds (mainly in lieu of Iraqi acceptance that the border in Shattol-arab or Arvandrood was Thalweg (the line dividing the deepest navigable channel of the river). The Iraqi position until was that the Iranian side of Shattol-arab was the borderline and even if the Iranians set foot to the waters of the river, they had entered the territory of Iraq.

Kurds considered the treaty as a betrayal to their cause and a treaty that stopped them form getting their goals in the region. Immediately after the 1975 treaty, Iran stopped supporting the Kurds that the armed forces of Saddam crushed the Kurdish rebellion in a serious way. The regime of Iraq even relocated some of the Kurdish population to change the demographic characteristics of the traditionally Kurdish areas. In fact, Talebai, the Iraqi president had referred to this point in his reference to the 1975 Treaty saying that al opposition groups at the time of Saddam opposed this treaty and considered it as the move that saved Saddam.

New Dimensions

Since the collapse of Saddam Hussein in Iraq, the Iranian Islamic Republic feels that there is a serious role for it to play in Iraq. The Iranian regime has interfered in the Iraqi affairs in many ways and this is well recognized. Recently, the regime of Iran has started to a new kind of intervention in the north of Iraq, the Kurdish section of Iraq (in collaboration with Turkish officials). The Iranian forces have bombarded the Kurdish villages in certain parts of Iraqi Kurdish region under the pretext of fighting the PKK elements.

Apart from the historical points, it is possible, that referring to the 1975 Treaty by the Iraqi president has an interesting side too. As it is known the treaties on the basis of the international law principles are subject to the succession of states. So if the Iraqis can force the Iranian say seriously that the 1975 Treaty is very valid, then they may ask Iranians either to abide by the treaty that they love it so dearly or see that the whole thing will go down.

Aside from the section of the treaty that talks about the borders in the Shattoarab, there are important articles in the treaty that makes the Iranian side committed to stop any interference in the internal affairs of Iraq. It says that Iran has no right to interfere in the internal affairs of Iraq, and if does, the other parts of the treaty, including the border arrangement are under question. (6)

Notes:

(1) Report of Al-hayat Arabic Newspaper, published in London, 18 December 2007.

(2) http://www.presstv/pop/print.aspx?id=36431 dated 26 Dec. 2007)

(3) http://www.radofarda.com/Article/2007/12/29/ f8_Talebani_Aligiers

(4) http://en.wikipedia.org/wiki/Algiers_Agreement_ (1975)

(5) http://www.danielpipes.org/pf.php?id=164

(6) International Law of the Sea and problems of Iran, by Bahman Aghai Diba, In Persian, the Legal regime of Arvand Rood, Page 126-135, Tehran, Ghangedanesh Publications, 1374 (Iranian Year). In this regard, it is important to note that in the US- Iran Arbitration Tribunal that was in fact formed by another "Algiers Accord" (this was intentionally called Algiers Statement? to avoid giving the impression that the two sides were actually entering into some kind of treaty but as a matter of fact, the Algiers Statement was a treaty)following the hostage taking in the American embassy in Tehran, has heard numerous cases by the two sides that have referred to the "1955 Treaty of Amity of Iran and USA".

(http://www.legislation.gov.hk/doc/multi_904v1.pdf)

Iran-Afghanistan: water dispute over Hirmand or Helmand River

Hirmand or Helmand River in the borders of Iran and Afghanistan is a potential point of conflict on scarce water resources of the region between the two countries. It has a long history of agreements and conflicts. Iran will most probably make its assistance to the landlocked state of Afghanistan dependent of cooperation of Afghan Government on the water share of Iran from Helmand.

In 1969-1973, Iranian diplomats were engaged in intense negotiations with Afghan counterparts over the waters of Helmand. Iran had the upper hand and, legally, was in a position to push for a fair share of the water. Disregarding the ongoing negotiations, however, the Shah of Iran, prompted by the sensitive political situation in Afghanistan, decided to grant a major concession to the Afghans. Immediately after the political decision, Iran and Afghanistan signed the 1973 treaty on the Helmand River, thus formalizing Iran's unequal share of the waters.

Iranian Minister of Court Asadollah Alam wrote in his diaries in March 1969 that Kabul would agree to ensure water flow to Iran only in exchange for credit facilities, improved access to Iranian ports, and development assistance ("The Shah and I," Alinaghi Alikhani, ed., London, 1991). And, when the Afghan legislature discussed a new agreement on the Hirmand River in October 1972, Tehran feared that it would be costly. Iranian monarch Mohammad Reza Pahlavi said, "authorize [the Iranian ambassador to Kabul] to make the pay offs if you really think they're necessary." Iranian Prime Minister Amir Abbas Hoveida and Afghan Prime Minister Mohammad Musa Shafiq signed an accord in 1973 that determined the specific amount of water that should flow into Iran: 26 cubic meters of water per second. Yet this agreement was not ratified. (http://www.payvand.com/news/05/sep/1061.html, 9/8/2005)

Relations between the Iranian government and the Taliban regime were troubled, and the two sides did not reach an accord on use of the River's water. During the Taliban rule (1994-2001) the water of Helmand or Hirmand to Iran was completely cut.

The Taliban's ouster and friendly relations between the governments of presidents Hamid Karzai and Hojatoleslam Mohammad Khatami suggested that the situation would change for the better. This would be a welcome development, as a multiyear drought had seriously affected the entire region. "The least we expect is implementation of the accord signed between Iran and Afghanistan before the Islamic revolution in Iran," parliamentarian Alaedin Borujerdi said on 1 September 2002, IRNA reported. (http://www.payvand.com/news/05/sep/1061.html, 9/8/2005)

The parliamentarian Gholam Hussein Aghai, who represented the Sistan va Baluchistan Province city of Zabol, also decried the Afghans' failure to provide water despite a new agreement reached during President Khatami's 13 August 2002 visit to Afghanistan, "Entekhab" reported on 1 September 2002. (http://www.payvand.com/news/05/sep/1061.html, 9/8/2005)

In September 2004, Iranian and Afghan officials met in Tehran for a joint meeting within the framework of the 1973 Helmand River treaty. Deputy Energy Minister Reza Ardakanian told IRNA on 8 September 2004 that the two sides were preparing for the implementation of the treaty. He said that under normal circumstances, Iran's annual share is 820 million cubic meters. (http://www.payvand.com/news/05/sep/1061.html, 9/8/2005)

The issue is still pending and has potential to turn into a more serious conflict due to the pressure that Iran feels in the Sistan and Baluchistan area.

In 2009, Shariyari the Representative of Zahedan in the Islamic Assembly of Iran [Majles] has said: "it is close to one month that the Government of Afghanistan has stopped the water of Helmand from reaching Iran and the agricultural activities around Zabol [in Iran's Baluchistan and Sistan Province] is seriously damaged... In 1973 Teary of Iran and Afghanistan on Helmand, it is provided that: annually 850 million cubic meters of the Helmand belong to Iran. However, Afghans believe that Iran's water share includes the spring floods, but Iran rejects this. The most important issue is the lack of a state control on the Afghan side over the course of Helmand River. Therefore, irresponsible elements are making channels and stealing the water of Helmand for their own purposes. The water of Kajaki Dam is used to irrigate the poppy plantations that are out of the supervision and control of the Afghan government. " (Tabnak , http://www.tabnak.ir/pages/print.php?cid=35168 dated 2/2/2009 in an article under the title of " Iran's water share of Helmand goes to irrigation of poppy plantations in Afghanistan")

The Research Center of the Iranian Majles has reported that all treaties of Iran and Afghanistan regarding the waters of Helmand River has been concluded contrary to the international regulations on the international rivers and asked for revision of them in the light of new realities.

The report says: "at the moment the 1351 [193] Treaty is valid and although the treaty has many shortcomings and generally it is arranged in favor of the Afghans and against the rights of Iran, the Afghan officials do not implement it and Iran is not actually get-

ting the meager rights it has under the treaty." http://www.tabnak.
ir/pages/print.php?cid=35168

According to Hassan Ghashqavi, the Spokesperson for the Iranian Ministry of Foreign affairs: " the Tenth Meeting of the Helmand Waters Commissioners was held in Kabul (2009) and the Iranian delegation protested strongly to the stopping of the water from Helmand and asked the Afghan Government to act to its commitments on the basis of the 1973 Treaty of Helmand....he added that the issue of Iran's share from Helmand will remain in the agenda of the Iranian Ministry of Foreign Affairs." http://
www.tabnak.ir/pages/print.php?cid=35168

Iran is committed and interested in helping the development process in Afghanistan. Iran has taken part and made commitments for helping Afghanistan. The most recent case was participation of Iran in the Hague Conference in 2009. The roads and transportation system of Iran is the best way of helping the land locked Afghanistan.

However, the serious sufferings of the people of Iran in the Sistan Province due to the shortage of water that has historically flowed to their lands can create serious obstacle in the relations. As the water conflicts are getting more serious in the region (and in many other places), the Iranian government may find it difficult to help the Afghan Government in the developmental projects while they deny what the Iranian side considers as its fair share of the Helmand River.

Iranian Baluchistan is also one of Iran's poorest and most underserved provinces. Tehran has great difficulty administering law and order in the region, having to rely instead on harsh security crackdowns that alienate the public. Given its poverty, lawlessness, and porous border with Pakistan, Iranian Baluchistan has emerged as a smuggler's paradise, a reputation that has made it both a regular target of the Iranian security services and an attractive base for enterprising criminals. (http://www.countercurrents.org/zambelis230509.htm) The activities of Jundollah group (Army of God) according to Iran, is "an extremist Sunni/Wahabi terrorist organization with links to Al-Qaeda, based in Baluchistan of Pakistan." Iran has regularly accused Jundollah of receiving support from US, British, and Saudi intelligence in an effort to destabilize the Islamic Republic from within by fomenting ethnic and sectarian strife) in the area, is basically benefiting from bad economic conditions of the people dependent on agricultural work in Baluchistan and the flow of Helmand water to the area is getting more political dimensions for both sides.

Apparently the Iranian Government insists that the provisions of the 1973 Treaty on Helmand can the basis of the cooperation between the two sides and the Afghan side believes that the treaty was not ratified by the Parliament of Afghanistan and it is null and void. Treaty was priding that 26 cubic meters of water per second should flow to Iran. The treaty was in fact concluded by the mediation efforts of the US and its negotiations took place with the US help.

In this case of the annulment of the treaty, the situation should return to the Status Quo Ante or the exiting situation before the treaty. Also, the two sides have formed joint commissions to look into the water conflicts and they have made the 1973 the basis of their efforts.

According to the International Law on the legal regime of the international rivers the riparian states have clear rights that should be observed. Upper riparian states are under an obligation not to prevent such waters from flowing to lower riparian country. (http://books.google.com/books?id=imnLhNiRfPU C&pg=PA71&lpg=PA71&dq=1973+Helmand+River+Treaty&sou rce=bl&ots=pCA2RA2Rm2&sig=8r7rqztmD7WSXH-WZbTnEfT CV5o&hl=en&ei=nzklSpfmKNGptgeYnanmBg&sa=X&oi=book result&ct=result&resnum=5)

Thanks to the financial institutions that act as the specialized agencies of the UN (especially the International Monetary Fund and World Bank) the international law on the international rivers has improved. The reason for serious involvement of such agencies in the concerned subject is that they do not give loans to states who intend to violate the riparian rights and when they give the loan make it conditional on observation of those rights.

The Afghan side believes Iran has taken the instability of Afghanistan for getting more of the Helmand water. They believe Iran has made several channels from the Helmand, while according to the 1973 Treaty they could have two channels only. The Afghan side believes that Iran is depositing the waters near Zabol. (http://www.rahenejatdaily.com/1007/87080912.html). Also, the 1973 treaty provided that Iran "buys" some of the waters of Helmand from Afghanistan and the Afghan side believes that Iran is not paying for those parts.

It seems that both Iran and Afghanistan look at the 1973 Treaty as something imposed on them. At the same time, they have acted as if the treaty can be "saved".

According to Hirmand (Helmand) River Water Treaty 1973 between Iran and Afghanistan and with regards to its Protocol No.1, both countries have participated Hirmand (Helmand) River common commissioners sessions since 2003. During above sessions two countries try to solve problems about Iran water rights based on monthly table of water distribution (Protocol No.2) and also other common activities in the river.

(http://portal.worldwaterforum5.org/wwf5/en-us/Lists/ Contributions/DispForm_Custom.aspx?List=07b1928d-1369-4ceb-9bb2-9201fcfaaca2&ID=445)

Attachments:

Iran Maritime Laws:

Act on the Marine Areas of the Islamic Republic of Iran in the Persian Gulf and the Oman Sea, 1993

Article 1
Sovereignty
The sovereignty of the Islamic Republic of Iran extends, beyond its land territory, internal waters and its islands in the Persian Gulf, the strait of Hormuz and the Oman Sea, to a belt of sea, adjacent to the baseline, described as the territorial sea.

This sovereignty extends to the airspace over the territorial sea as well as to its bed and subsoil.

Article 2
Outer limit
The breadth of the territorial sea is 12 nautical miles, measured from the baseline. Each nautical mile is equal to 1,852 meters. The islands belonging to Iran, whether situated within or outside its territorial sea, have, in accordance with this Act, their own territorial sea.

Article 3
Baseline
In the Persian Gulf and the Oman Sea, the baseline from which the breadth of the territorial sea is measured is that one determined in Decree No. 2/250-67 dated 31 Tir 1352 (22 July 1973) of the Council of Ministers (annexed to this Act); in other areas and islands, the low-water line along the coast constitutes the baseline.

Waters on the landward side of the baseline of the territorial sea, and waters between islands belonging to Iran, where the distance of such islands does not exceed 24 nautical miles, form part of the internal waters and are under the sovereignty of the Islamic Republic of Iran.

Article 4
Delimitation
Wherever the territorial sea of Iran overlaps the territorial seas of the States with opposite or adjacent coasts, the dividing line between the territorial seas of Iran and those states shall be, unless otherwise agreed between the

two parties, the median line every point of which is equidistant from the nearest point on the baseline of both States.

Page 2

National legislation - DOALOS/OLA - United Nations

Article 5
Innocent passage
The passage of foreign vessels, except as provided for in article 9, is subject to the principle of innocent

passage so long as it is not prejudicial to good order, peace and security of the Islamic Republic of Iran.

Passage, except as in cases of force majeure, shall be continuous and expeditious.

Article 6
Requirements of innocent passage

Passage of foreign vessels, in cases when they are engaged in any of the following activities, shall not be considered innocent and shall be subject to relevant civil and criminal laws and regulations:

(a) Any threat or use of force against the sovereignty, territorial integrity or political independence of the Islamic Republic of Iran, or in any other manner in violation of the principles of international law;

(b) Any exercise or practice with weapons of any kind;

(c) Any act aimed at collecting information prejudicial to the national security, defense or economic interests of the Islamic Republic of Iran;

(d) Any act of propaganda aimed at affecting the national security, defense or economic interests of the Islamic Republic of Iran;

(e) The launching, landing or transferring on board of any aircraft or helicopter, or any military devices or personnel to another vessel or to the coast;

(f) The loading or unloading of any commodity, currency or person contrary to the laws and regulations of the Islamic Republic of Iran;

(g) Any act of pollution of the marine environment contrary to the rules and regulations of the Islamic Republic of Iran;

(h) Any act of fishing or exploitation of the marine resources;

(i) The carrying out of any scientific research and cartographic and seismic surveys or sampling activities;

(j) Interfering with any systems of communication or any other facilities or installations of the Islamic Republic of Iran;

(k) Any other activity not having a direct bearing on passage.

Page 3

National legislation - DOALOS/OLA - United Nations asdf

Article 7
Supplementary laws and regulations
The Government of the Islamic Republic of Iran shall adopt such other regulations as are necessary for the
protection of its national interests and the proper conduct of innocent passage.

Article 8
Suspension of innocent passage
The Government of the Islamic Republic of Iran, inspired by its high national interests and to defend its security, may suspend the innocent passage in parts of its territorial sea.

Article 9
Exceptions to innocent passage
Passage of warships, submarines, nuclear-powered ships and vessels or any other floating objects or vessels carrying nuclear or other dangerous or noxious substances harmful to the environment, through the territorial sea is subject to the prior authorization of the relevant authorities of the Islamic Republic of Iran. Submarines are required to navigate on the surface and to show their flag.

Article 10
Criminal jurisdiction
In the following cases, the investigation, prosecution and punishment in connection with any crimes committed on board the ships passing through the territorial sea is within the jurisdiction of the judicial authorities of the Islamic Republic of Iran:
- (a) If the consequences of the crime extend to the Islamic Republic of Iran;
- (b) If the crime is of a kind to disturb the peace and order of the country or the public order of the territorial sea;
- (c) If the master of the ship or a diplomatic agent or consular officer of the flag State asks for the assistance and investigation;

(d) If such investigation and prosecution is essential for the suppression of illicit traffic in narcotic drugs or psychotropic substances.

Article 11
Civil jurisdiction

The competent authorities of the Islamic Republic of Iran may stop, divert or detain a ship and its crew for the enforcement of attachment orders or court judgments if:

(a) The ship is passing through the territorial sea after leaving the internal waters of Iran;

(b) The ship is lying in the territorial sea of Iran;

Page 4

National legislation - DOALOS/OLA - United Nations asdf

(c) The ship is passing through the territorial sea, provided that the origin of the attachment order or court judgment rests in the obligations or requirements arising from the civil liability of the ship itself.

PART II
Contiguous zone

Article 12
Definition

The contiguous zone is an area adjacent to the territorial sea the outer limit of which is 24 nautical miles from the baseline.

Article 13
Civil and criminal jurisdiction

The Government of the Islamic Republic of Iran my adopt measures necessary to prevent the infringement of laws and regulations in the contiguous zone, including security, customs, maritime, fiscal, immigration, sanitary and environmental laws and regulations and investigation and punishment of offenders.

PART III
Exclusive economic zone and continental shelf

Article 14
Sovereign rights and jurisdiction in the exclusive economic zone

Beyond its territorial sea, which is called the exclusive economic zone, the Islamic Republic of Iran exercises its sovereign rights and jurisdiction with regard to:

(a) Exploration, exploitation, conservation and management of all natural resources, whether living or non-living, of the seabed and subsoil thereof and its superjacent waters, and with regard to other economic activities for the production of energy from water, currents and winds. These rights are exclusive;

(b) Adoption and enforcement of appropriate laws and regulations, especially for the following activities:

(i) The establishment and use of artificial islands and other installations and structures, laying of submarine cables and pipelines and the establishment of relevant security and safety zones;

(ii) Any kind of research;

Page 5

National legislation - DOALOS/OLA - United Nations asdf

(iii) The protection and preservation of the marine environment;

(c) Such sovereign rights as granted by regional or international treaties.

Article 15
Sovereign rights and jurisdiction in the continental shelf

The provisions of article 14 shall apply mutatis mutandis to the sovereign rights and jurisdiction of the Islamic Republic of Iran in its continental shelf, which comprises the seabed and subsoil of the marine areas that extend beyond the territorial sea throughout the natural prolongation of the land territory.

Article 16
Prohibited activities
Foreign military activities and practices, collection of information and any other activity inconsistent with the rights and interests of the Islamic Republic of Iran in the exclusive economic zone and the continental shelf are prohibited.

Article 17
Scientific activities, exploration and research
Any activity to recover drowned objects and scientific research and exploration in the exclusive economic zone and the continental shelf is subject to the permission of the relevant authorities of the Islamic Republic of Iran.

Article 18
Preservation of the environment and natural resources
The Government of the Islamic Republic of Iran shall take appropriate measures for the protection and preservation of the marine environment and proper exploitation of living and other resources of the exclusive economic zone and the continental shelf.

Article 19
Delimitation
The limits of the exclusive economic zone and the continental shelf of the Islamic Republic of Iran, unless otherwise determined in accordance with bilateral agreements, shall be a line every point of which is equidistant from the nearest point on the baselines of two States.

Article 20
Civil and criminal jurisdiction
The Islamic Republic of Iran shall exercise its criminal and civil jurisdiction against offenders of the laws and regulations in

the exclusive economic zone and continental shelf and shall, as appropriate, investigate or detain them.

Page 6

National legislation - DOALOS/OLA - United Nations asdf

Article 21
Right of hot pursuit

The Government of the Islamic Republic of Iran reserves its right of hot pursuit against offenders of laws and regulations relating to its internal waters, territorial sea, contiguous zone, exclusive economic zone and the continental shelf, in such areas and the high seas.

PART IV
Final provisions

Article 22
Executive regulations

The Council of Ministers shall specify the mandates and responsibilities [powers and duties] of different ministries and organizations charged with the enforcement of this Act.

The said ministries and organizations shall, within one year after the approval of this Act, prepare the necessary regulations and have them approved by the Council of Ministers.

Pending the adoption of new executive regulations, the existing rules and regulations shall remain in force.

Article 23

All laws and regulations contrary to the present Act, upon its ratification, are hereby abrogated.

The above Act, comprising 23 articles, was ratified at the plenary meeting of Tuesday, the thirty-first day of Farvrdin, one thousand three hundred and seventy-two (20 April 1993), of the Islamic Consultative Assembly and was approved by the Council of Guardians on Ordibehesht 12, 1372

(2 May 1993).

Iran Act dated 21 July 1973 (the baselines)

(http://www.un.org/Depts/los/LEGISLATION-
ANDTREATIES/PDFFILES/IRN_1973_Decree.pdf)

The baseline, established in the Act of 22 Farvardin 1338 (12 April 1959) amending the Act of 24 Tir 1313
(15 July 1934) concerning the limits of the territorial waters and the contiguous zone of Iran, is determined as follows: A. Straight lines joining the following points:

(1) Point 1, situated at the point where the thalweg of the Shatt/ El-Arab intersects the straight line joining the two banks of the mouth of the Shatt El-Arab at the low-water line.

(2) Point 2, situated at the mouth of the Behregan, whose geographical co-ordinates are: latitude 29°59'50" N and longitude 49°33'55" E.

(3) Point 3, situated on the south coast of Kharg Island, whose geographical co-ordinates are: latitude 29°12'29" N and longitude 50°18'40" E.

(4) Point 4, situated on the south coast of Nakhilu Island, whose geographical co-ordinates are: latitude 27°50'40" N and longitude 51°27'15" E.

(5) Point 5, situated on Lavan Island, whose geographical co-ordinates are: latitude 26°47'25" N and longitude 53°13'00" E.

(6) Point 6, situated on the south-west coast of Kish Island, whose geographical co-ordinates are: latitude 26°30'55" N and longitude 53°55'10" E.

(7) Point 7, situated on the south-east coast of Kish Island, whose geographical co-ordinates are: latitude 26°30'10" N and longitude 53°59'20" E.

(8) Point 8, situated at Ras-o-Shenas, whose geographical co-ordinates are: latitude 26°29'35" N and longitude 54°47'20"E.

(9) Point 9, situated on the south-west coast of Qeshm Island, whose geographical co-ordinates are: latitude 26°32'25" N and longitude 55°60'55" E.

(10) Point 10, situated on the south coast of Hengam Island, whose geographical co-ordinates are: latitude 26°36'40" N and longitude 55°51'50" E.

(11) Point 11, situated on the south coast of Larak Island, whose geographical co-ordinates are: latitude 26°49'30" N and longitude 56°21'50" E.

(12) Point 12, situated on the east coast of Larak Island, whose geographical co-ordinates are: latitude 26°51'15" N and longitude 56°24'05" E.

(13) Point 13, situated on the east coast of Hormoz Island, whose geographical co-ordinates are: latitude Page 2 National legislation - DOALOS/OLA - United Nations asdf 27°02'30" N and longitude 56°29'40" E.

(14) Point 14, whose geographical co-ordinates are: latitude 27°08'30" N and longigude 56°35'40" E.

(15) Point 15, whose geographical co-ordinates are: latitude 25°47'10" N and longitude 57°19'55" E.

(16) Point 16, whose geographical co-ordinates are: latitude 25°38'10" N and longitude 57°45'30" E.

(17) Point 17, whose geographical co-ordinates are: latitude 25°33'20" N and longitude 58°05'20" E.

(18) Point 18, whose geographical co-ordinates are: latitude 25°24'05" N and longitude 59°05'40" E.

(19) Point 19, whose geographical co-ordinates are: latitude 25°23'45" N and longitude 59°35'00" E.

(20) Point 20, whose geographical co-ordinates are: latitude 25°19'20" N and longitude 60°12'10" E.

(21) Point 21, whose geographical co-ordinates are: latitude 25°17'25" N and longitude 60°24'50" E.

(22) Point 22, whose geographical co-ordinates are: latitude 25°16'36" N and longitude 60°27'30" E.

(23) Point 23, whose geographical co-ordinates are: latitude 25°16'20" N and longitude 60°36'40" E.

(24) Point 24, whose geographical co-ordinates are: latitude 25°03'30" N and longitude 61°25'00" E.

(25) Point 25, situated at the point of intersection of the meridian 61°37'03" E and the straight line joining the shorelines at the entrance of the Gwadar Gulf at the low-water line.

B. Between points 6 and 7, situated on Kish Island, points 11 and 12, situated on Larak Island, and points 14 and 15, situated in the Strait of Hormuz, the low-water line shall constitute the baseline.

II.

The baseline used for measuring the breadth of the territorial sea of Iran is shown on the Map of the Persian Gulf, the first edition of which was published in Shahrivar 1349 (September 1970) by the National Geographical Organization of Iran, on a scale of 1:1,500,000, and is attached to the present Decree. The original of the Decree is kept in the Office of the President of the Council of Ministers.

Iran Iraq Treaty of 1975 and protocols

His Imperial Majesty the Shahanshah of Iran,
His Excellency the President of the Republic of Iraq,

Considering the sincere desire of the two Parties as expressed in the Algiers Agreement of 6 March 1975, to achieve a final and lasting solution to all the problems pending between the two countries,

Considering that the two Parties have carried out the definite re-demarcation of their land frontier on the basis of the Constantinople Protocol of 1913 and the minutes of the meetings of the frontier Delimitation Commission of 1914 and have delimited their river frontier along the thalweg,

Considering their desire to restore security and mutual trust throughout the length of their common frontier,

Considering the ties of geographical proximity, history, religion, culture and civilization which bind the peoples of Iran and Iraq,

Desirous of strengthening their bonds of friendship and neighborliness, expanding their economic and cultural relations and promoting exchange and human relations between their peoples on the basis of the principles of territorial integrity, the inviolability of frontiers and non-interference in internal affairs,

Resolved to work towards the introduction of a new era in friendly relations between Iran and Iraq based on full respect for the national independence and sovereign equality of states,

Convinced that they are helping thereby to implement the principles and achieve the purposes and objectives of the Charter of the United Nations,

Have decided to conclude this Treaty and have appointed as their plenipotentiaries:

His Imperial Majesty the Shahinshah of Iran:

His Excellency Abbas Ali Khalatbary, Minister of Foreign Affairs of Iran.

His Excellency the President of Iraq:

His Excellency Saadoun Hamadi, Minister for Foreign Affairs of Iraq.

Who, having exchanged their full powers, found to be in good and due form, have agreed as follows:

Article 1

The High Contracting Parties confirm that the State land frontier between Iraq and Iran shall be that which has been re-demarcated on the basis of and in accordance with the provisions of the Protocol concerning the re-demarcation of the land frontier, and the annexes thereto, attached to this Treaty.

Article 2

The High Contracting Parties confirm that the State frontier in the Shatt Al Arab shall be that which has been delimited on the basis of and in accordance with the provisions of the Protocol concerning the delimitation of the river frontier, and the annexes thereto, attached to this Treaty.

Article 3

The High Contracting Parties undertake to exercise strict and effective permanent control over the frontier in order to put an end to any infiltration of a subversive nature from any sources, on the basis of and in accordance with the provisions of the protocol concerning frontier security, and the annex thereto, attached to this Treaty.

Article 4

The High Contracting Parties confirm that the provisions of the three Protocols, and the annexes thereto, referred to in article 1, 2, and 3 above and attached to this Treaty as an integral part thereof shall be final and permanent. They shall not be infringed under any circumstances and shall constitute the indivisible elements of an over-all settlement. Accordingly, a breach of any of the components of this over-all settlement shall clearly be incompatible with the spirit of the Algiers Agreement.

Article 5

In keeping with the inviolability of the frontiers of the two States and strict respect for their territorial integrity, the High Contracting Parties confirm that the course of their land and river frontiers shall be inviolable, permanent and final.

Article 6

1. In the event of a dispute regarding the interpretation or implementation of this Treaty, the three Protocols or the annexes thereto, any solution to such a dispute shall strictly respect the course of the Iraqi-Iranian frontier referred to in articles 1 and 2 above, and shall take into account the need to maintain security on the Iraqi-Iranian frontier in accordance with article 3 above.

2. Such disputes shall be resolved in the first instance by the High Contracting Parties, by means of direct bilateral negotiations to be held within two months after the date on which one of the Parties so requested.

3. If no agreement is reached, the High Contracting Parties shall have recourse, within a three-month period, to the good offices of a friendly third State.

4. Should one of the two Parties refuse to have resource to good offices or should the good offices procedure fail, the dispute shall be settled by arbitration within a period of not more than one month after the date of such refusal or failure.

5. Should the High Contracting Parties disagree as to the arbitration procedure, one of the High Contracting Parties may have recourse, within 15 days after such disagreement was recorded, to a court of arbitration.

With a view to establish such a court of arbitration each of the High Contracting Parties shall, in respect of each dispute to be resolved, appoint one of its nationals as arbitrators and the two arbitrators shall choose an umpire. Should the High Contracting Parties fail to appoint their arbitrators within one month after the date on which one of the Parties received a request for arbitration from the other Party, or should the arbitrators fail to reach agreement on the choice of the umpire before that time-limit expires, the High Contracting Party which requested arbitration shall be entitled to request the President of the International Court of Justice to appoint the arbitrators or the umpire, in accordance with the procedures of the Permanent Court of Arbitration.

6. The decision of the court of arbitration shall be binding on and enforceable by the High Contracting Parties.

7. The High Contracting Parties shall each defray half the costs of the arbitration.

Article 7
This Treaty, the three Protocols and the annexes thereto shall be registered in accordance with Article 102 of the Charter of the United Nations.

Article 8

This Treaty, the three Protocols and the annexes thereto shall be ratified by each of the High Contracting Parties in accordance with its domestic law.

This Treaty, the three Protocols and the annexes thereto shall enter into force on the date of the exchange of the instruments of ratification in Tehran.

IN WITNESS WHEREOF the Plenipotentiaries of the High Contracting Parties have signed this Treaty, the three Protocols and the annexes thereto.

DONE at Baghdad, on 13 June 1975.

(Signed) (Signed)

Abbas Ali Khalatbary Saadoun Hamadi

Minister for Foreign Minister for Foreign

Affairs of Iran Affairs of Iraq

This Treaty, the three Protocols and the annexes thereto were signed in the presence of His Excellency Abdel-Aziz Bouteflika, Member of the Council of the Revolution and Minister for Foreign Affairs of Algeria.

(Signed)

Protocol concerning the Re-demarcation of the Land Frontier between Iran and Iraq

Pursuant to the provisions of the Algiers communiqué of 5 March 1975,

The two Contracting Parties have agreed as follows:

Article 1

A. The two Contracting Parties affirm and recognize that the re-demarcation of the State land frontier between Iran and Iraq was a field operation performed by the mixed Iraqi-Iranian-Algerian Committee on the basis of the following:

1. The Constantinople Protocol of 1913 and the minutes of the meetings of the 1914 Commission to delimit the Turco-Persian frontier;

2. The Tehran Protocol dated 17 March 1975;

3. The record of the meeting of Ministers for Foreign Affairs, signed at Baghdad on 20 April, 1975 and approving, inter alia, the record of the Committee to Demarcate the Land frontier, signed at Tehran on 30 March, 1975;

4. The record of the meeting of Ministers for Foreign Affairs, signed at Algiers on 20 May, 1975;

5. The descriptive record of operations in the demarcation of the land frontier between Iran and Iraq, prepared by the Committee to Demarcate the Land Frontier and dated 13 June, 1975. The record constitutes Annex 1 and is an integral part of this Protocol;

6. Maps on the scale 1:50,000 indicating the land frontier line and the position of the old and new frontier marks. The maps constitute Annex 2 and are in integral part of this Protocol.

7. Record cards of the old and new frontier marks;

8. A document giving the co-ordinates of the frontier marks;

9. Aerial photographs of the Iraqi-Iranian frontier strip indicating the positions of the old and new frontier marks.

B. The two Parties undertake to complete the demarcation of the frontier between frontier marks No. 14 and No. 15 within two months.

C. The two Contracting Parties shall co-operate in producing aerial photographs of the Iranian-Iraqi land frontier with a view to using them in plotting the frontier on maps on the scale 1:25,000, indicating the position of the frontier marks. This work shall be completed within a period not exceeding one year with effect from 20 May, 1975, and shall be without prejudice to the entry into force of the Treaty of which this Protocol is an integral part.

The descriptive record relating to the land frontier and referred to in paragraph 5 above shall be amended accordingly.

The maps produced pursuant to the present section C shall supersede all existing maps.

Article 2

The State land frontier between Iraq and Iran shall follow the line indicated in the descriptive record and the maps referred to respectively in paragraphs 5 and 6 of Article 1 above, due regard being had to the provisions of section C of that Article.

Article 3

The frontier line defined in Articles 1 and 2 of this Protocol shall also divide vertically the air space and the subsoil.

Article 4

The two Contracting Parties shall established a Mixed Iraqi-Iranian Commission to settle, in a neighborly and co-operative spirit, the status of landed property, constructions, or technical or other installations whose national character may be changed by the re-demarcation of the land frontier between Iraq and Iran. Such settlement shall be by means of repurchase compensation or any other appropriate formula, with a view to eliminating any source of litigation.

The Commission shall settle the status of State property within two months. Claims concerning private property shall be submitted to it within two months. The status of this private property shall be settled within the following there months.

Article 5

1. A Mixed Commission composed of representatives of the competent authorities of the two States shall be established to inspect the frontier marks and determine their condition.

2. Either Contracting Party may request the other in writing to have the Commission carry out, at any time, an additional inspection of the frontier marks. In the event of such a request, the inspection shall be made within a period not exceeding 30 days after the date of the request.

3. Whenever an inspection is made, the Mixed Commission shall prepare the relevant reports and submit them under its sig-

nature to the competent authorities of each of the two States. The Commission may, if need be, call for the construction of new frontier marks according to the specifications of the existing ones provided that the course of the frontier line is not thereby altered. Where new frontier marks are constructed, the competent authorities of the two States shall check the frontier marks and their coordinates against the relevant maps and documents referred to in Article 1 of this Protocol. The authorities shall then position the frontier marks under the supervision of the Mixed Commission, which shall prepare a record of the operation and submit it to the competent authorities of each of the two States so that it may be annexed to the documents referred to in Article 1 of this Protocol.

4. The two Contracting Parties shall be jointly responsible for the maintenance of the frontier marks.

5. The Mixed Commission shall be responsible for replacing displaced frontier marks and reconstructing destroyed or missing marks, on the basis of this Protocol, taking care not to alter, under any circumstances, the position of the marks. In such cases, the Mixed Commission shall prepare a record of the operation and submit it to the competent authorities of each of the two States.

6. The competent authorities of each of the two States shall exchange information on the condition of the frontier marks with a view to finding the best ways and means of protecting and maintaining them.

7. The two Contracting Parties undertake to take all necessary steps to protect the frontier marks and prosecute individuals who have moved, damaged or destroyed them.

Article 6

The two Contracting Parties have agreed that the provisions of this Protocol, signed without any reservation, shall henceforth govern any matter relating to the frontier between Iran and Iraq. On these basis, they solemnly undertake to respect their common and definitive frontier.

Done at Baghdad, on 26 December, 1975
(Signed) (Signed)
Abbas Ali Khalatbary Saadoon Hammadi
Minister of Foreign Affairs Minister of Foreign Affairs
of Iran of Iraq

Signed in the presence of His Excellency Abdel-Aziz Boutef-like, Member of the Council of the Revolution, Minister for Foreign Affairs of Algiera.

Record

In connection with the description of the Iranian-Iraqi land frontier annexed to the Protocol concerning the redemarcation of the land frontier between Iran and Iraq, of 13 June 1975, the undersigned representatives of Iran and Iraq, duly empowered, have reached agreement on the following arrangements:

1. With regard to the description of the course of the frontier between mark No. 101 and Mark No. 101/1, it has been agreed that the frontier line shall run between two springs bearing the same name "Chiftekan".

Accordingly, the description set out in Annex 1 to the Protocol concerning the re-demarcation of the land frontier, which reads:

"It shall go in a straight line to a point situated between two springs bearing the same name "Chiftekan", whence it shall ascend in a straight line to the crest of Sour Kuh mountain",

signifies that the frontier line shall run between the two springs, which lie approximately 5 meters apart.

The existing arrangements for sharing the waters of the two springs (12 hours for Iran and 12 hours for Iraq in any 24 hours period) shall continue to apply.

The representatives of the two Parties deem it desirable to erect one or two additional marks between mark No. 101 and mark No. 101/1, in order to delineate more clearly the course of the frontier.

2. The description of the frontier between mark No. 81 and mark No. 82 set out in Annex 1 to the Protocol concerning the re-demarcation of the land frontier between Iran and Iraq, which reads:

"It shall then climb the Dere-i Tekkiyeh ravine, skirting the orchards situated therein in such a way as to leave them in Persian territory. From the point where these orchards end, it shall follow the Thalweg of the ravine..."

specifically provides that the frontier shall skirt all existing orchards, thus, leaving them in Iranian territory. The representatives of the two Parties have agreed on the creation of six additional marks in order to delineate more clearly the course of the frontier.

3. The erection of the additional marks referred to in paragraphs 1 and 2 above shall be effected in situ by the duly authorized representatives of the two countries. The marks shall be constructed as soon as weather conditions permit.

DONE at Baghdad, on 26 December, 1975

For Iran: For Iraq:

(Signed) (Signed)

General Ebrahim Khalvati Alladin Al-Sakkal

Protocol concerning the delimitation to the River Frontier between Iran and Iraq

Pursuant to the decisions taken in the Algiers communiqué of 6 March 1975

The two Contracting Parties have agreed as follows:

Article 1

The two Contracting Parties hereby declare and recognize that the State river frontier between Iran and Iraq in the Shatt-al-Arab has been delimited along the Thalweg by the Mixed Iraqi-Iranian-Algerian Committee on the basis of the following:

1. The Teheran Protocol of 17 March 1975;

2. The record of the Meeting of Ministers for Foreign Affairs signed at Baghdad on 20 April 1975, approving, inter alia, the record of the Committee to Delimit the River Frontier, signed on 16 April 1975, on the board of the Iraqi ship El Thawra in the Shatt-al-Arab.

3. Common hydrographic charts, which have been verified on the spot and corrected, and on which the geographical co-ordinates of the 1975 frontier crossing points have been indicated; these charts have been signed by the hydrographic experts of the Mixed Technical Commission and countersigned by the heads of the Iran, Iraq and Algerian delegations to the Committee. The said charts, listed hereinafter, are annexed to this Protocol and form an integral part thereof:

Chart No. 1: Entrance to the Shatt-al-Arab, No. 3842, published by the British Admiralty.

Chart No. 2: Inner Bar to Kabda Point, No. 3843, published by the British Admiralty.

Chart No. 3: Kabda Point to Abadan, No. 3844, published by the British Admiralty.

Chart No. 4: Abadan to Jazirat Ummat Tuwaylah, No. 3845, published by the British Admiralty.

Article 2

1. The frontier line in the Shatt-al-Arab shall follow the Thalweg, i.e., the median line of the main navigable channel at the lowest navigable level, starting from the point at which the land frontier between Iran and Iraq enters the Shatt-al-Arab and continuing to the sea.

2. The frontier line, as defined in paragraph 1 above, shall vary with changes brought about by natural causes in the main navigable channel. The frontier line shall not be affected by other changes unless the two Contracting Parties conclude a special agreement to that effect.

3. The occurrence of any of the changes referred to in paragraph 2 above shall be attested jointly by the competent technical authorities of the two Contracting Parties.

4. Any change in the bed of the Shatt-al-Arab brought about by natural causes which would involve a change in the national character of the two States, respective territory or of landed property, constructions, or technical or other installations shall not change

the course of the frontier line which shall continue to follow the Thalweg in accordance with the provisions of paragraph 1 above.

5. Unless an agreement is reached between the two Contracting Parties concerning the transfer of the frontier line to the new bed, the water shall be re-directed at the joint expense of both Parties to the bed existing in 1975 - as marked on the four common charts listed in Article 1, paragraph 3, above - should one of the Parties so require within two years after the date on which the occurrence of the change was attested by either of the two Parties. Until such time, both Parties shall retain their previous rights of navigation and of use over the water of the new bed.

Article 3

1. The river frontier between Iran and Iraq in the Shatt-al-Arab, as defined in Article 2 above, is represented by the relevant line drawn on the common charts referred to in Article 1, paragraph 3, above.

2. The two Contacting Parties have agreed to consider that the river frontier shall end at the straight line connecting the two banks of the Shatt-al-Arab, as its mouth, at the astronomical lowest low-water mark. This straight line has been indicated on the common hydrographic charts referred to in Article 1, paragraph 3, above.

Article 4

The frontier line as defined in Article 1, 2 and 3 of this protocol shall also divide vertically the air space and the subsoil.

Article 5

With a view to eliminating any sources of controversy, the two Contracting Powers shall established a Mixed Iraqi-Iranian Commission to settle, within two months, any questions concerning the status of landed property, constructions, technical or other installations, the national character of which may be affected by the delimitation of the Iranian-Iraqi river frontier, either through repurchase or compensation or any other suitable arrangement.

Article 6

Since the task of surveying the Shatt-al-Arab has been completed and the common hydrographic chart referred to in Article 1, paragraph 3, above has been drawn up, the two Contracting Parties have agreed that a new survey of the Shatt-al-Arab shall be carried out jointly, once every 10 years, with effect from the date of signature of this Protocol. However, each of the two Parties shall have the right to request new surveys, to be carried out jointly, before the expiry of the 10-years period.

Article 7

1. Merchant vessels, State vessels and warships of the two Contracting Parties shall enjoy freedom of navigation in the Shatt-al-Arab and in any part of the navigable channels in the territorial sea which lead to the mouth of the Shatt-al-Arab, irrespective of the line delimiting the territorial sea of each of the two countries.

2. Vessels of third countries used for purposes of trade shall enjoy freedom of navigation, on an equal and non-discriminatory basis, in the Shatt-al-Arab and in any part of the navigable cannels in the territorial sea which lead to the mouth of the Shatt-al-Arab, irrespective of the line delimiting the territorial sea of each of the two countries.

3. Either of the two Contracting Parties may authorize foreign warships visiting its ports to enter the Shatt-al-Arab, provided that such vessels do not belong to a country in a state of belligerency, armed conflict or war with either of the two Contracting Parties and provided that the other Party is so notified no less than 72 hours in advance.

4. The two Contracting Parties shall in every case refrain from authorizing the entry to the Shatt-al-Arab of merchant vessels belonging to a country in a state of belligerency, armed conflict or war with either of the two Parties.

Article 8

1. Rules governing navigation in the Shatt-al-Arab shall be drawn up by a mixed Iranian-Iraqi Commission, in accordance with the principles of equal rights of navigation for both States.

2. The two Contracting Parties shall establish a Commission to draw up rules governing the prevention and control of pollution in the Shatt-al-Arab.

3. The two Contracting Parties undertake to conclude subsequent agreements on the question referred to in paragraph 1 and 2 of this Article.

Article 9

The two Contracting Parties recognize that the Shatt-al-Arab is primarily an international waterway, and undertake to refrain from any operation that might hinder navigation in the Shatt-al-Arab or in any part of those navigable channels in the territorial sea of either of the two countries that lead to the mouth of the Shatt-al-Arab.

DONE at Baghdad, on 13 June 1975.

(Signed) (Signed)

Abbas Ali Khalatbary Saadoun Hamadi

Minister for Foreign Affairs Minister for Foreign Affairs

of Iran of Iraq

Signed in the presence of His Excellency Abdel Aziz Bouteflike, Member of the Council of the Revolution and Minister for Foreign Affairs of Algeria.

Protocol concerning Security on the Frontier between Iran and Iraq

In accordance with the decisions contained in the Algiers Agreement of 6 March 1975,

Anxious to re-establish mutual security and trust throughout the length of their common frontier,

Resolved to exercise strict and effective control over the frontier in order to put an end to any infiltration of a subversive nature, and, to that end, to establish close cooperation between themselves and to prevent any infiltration or illegal movement across their common frontier for the purpose of causing subversion, insubordination or rebellion,

Referring to the Teheran Protocol of 15 March 1975, the record of the meeting of Ministers for Foreign Affairs, signed at Baghdad on 20 April 1975, and the record of the meeting of the Ministers for Foreign Affairs, signed at Algiers on 20 May 1975.

The two Contracting Parties have agreed as follows:

Article 1

1. The two Contracting Parties shall exchange information of any movement by subversive elements who may attempt to infiltrate one of the two countries with a view to committing acts of subversion, insubordination or rebellion.

2. The two Contracting Parties shall take the necessary measures with regard to the movements of the elements referred to in paragraph 1 above.

The same steps shall be taken with regard to any persons who may assemble within the territory of one of the two Contracting Parties with the intention of committing acts of subversion or sabotage in the territory of the other Party.

Article 2

The many forms of co-operation established between the competent authorities of the two Contracting Parties relating to the closing of frontiers to prevent infiltration by subversive elements shall be instituted by the frontier authorities of the two countries and shall be pursued up to the highest levels in the Ministries of Defence, Foreign Affairs and the Interior of each of the two Parties.

Article 3

The infiltration points likely to be used by subversive elements are as follows:

Northern Frontier Zone

From the point of intersection of the Iranian, Turkish and Iraqi frontiers to (and including) Khanaqin-Qasr-e Shirin: 21 points.

Southern Frontier Zone

From (but not including) Khanaqin-Qasr-e Shirin to the end of the Iranian-Iraqi frontier: 17 points.

The above infiltration points are named in the annex.

The points specified above shall be supplemented by any infiltration point which may be discovered and will have to be closed and controlled.

5. All frontier crossing points except those currently controlled by the customs authorities shall be closed.

6.In the interests of promoting relations of all kinds between the two neighbouring countries, the two Contracting Parties have agreed that, in future, other crossing points controlled by the customs authorities shall be created by common consent.

Article 4

1. The two Contracting Parties undertake to provide the necessary human and material resources to ensure the effective closure and control of their frontiers, so as to prevent any infiltration by subversive elements through the crossing points mentioned in Article 3 above.

2. If, in the light of experience gained in this matter, experts should decide that more effective measures must be taken, the corresponding procedures shall be established at monthly meetings of the frontier authorities of the two countries, or at meetings between those authorities should the need arise.

The conclusions and records of such meetings shall be communicated to the higher authorities of each of the two Parties. Should there be disagreement between the frontier authorities, the heads of the administrations concerned shall meet in either Baghdad or Teheran to reconcile the points of view and draw up a record of the outcome of their meetings.

Article 5

1. Any subversive persons who may be arrested shall be handed over to the competent authorities of the Party in whose territory they were arrested and shall be subject to the legislation in force.

2. The two Contracting Parties shall inform one another of the measures taken against persons referred to in paragraph 1 above.

3. Should subversive persons cross the frontier in an attempt to escape, the authorities of the other country shall be informed immediately and shall take all necessary steps to apprehend such persons.

Article 6

In case of need where the two Contracting Parties so agree, entry to certain areas may be declared prohibited in order to prevent subversive persons from carrying out the intentions.

Article 7

In order to establish and promote co-operation which is mutually beneficial to both Parties, a permanent Mixed Committee comprising the heads of the frontier authorities and representatives of the Ministers for Foreign Affairs to the two countries shall be established and shall hold two sessions a year (at the beginning of each half of the calendar year).

At the request of one of the Parties, however, special meetings may be held to consider how intellectual and material resources might be better used for the closure and control of the frontiers and to review the effectiveness and proper implementation of the basic provisions governing co-operation as provided for in this Protocol.

Article 8

The provisions of this Protocol relating to the closure and control of the frontier shall be without prejudice to the provisions of specific agreements between Iran and Iraq concerning grazing rights and frontier commissioners.

Article 9

With a view to guaranteeing the security of the common river frontier in the Shatt-al-Arab and preventing the infiltration of subversive elements from either side, the two Contracting Parties shall take such appropriate steps as the installation of look-out posts and the detachment of patrol boats.

DONE at Baghdad, on 13 June 1975
(Signed) (Signed)
Abbas Ali Khalatbary Saadoun Hammadi
Minister for Foreign Affairs Minister for Foreign Affairs
of Iran of Iraq
Signed in the presence of H. E. Abdel Aziz Bouteflike, Minister for Foreign Affairs of Algeria.

INTERNATIONAL BOUNDARY STUDY
LIMITS IN THE SEAS
No. 114
IRAN'S MARITIME CLAIMS
March 16, 1994

This paper is one of a series issued by the Office of Ocean Affairs, Bureau of Oceans and International Environmental and Scientific Affairs in the Department of State.

IRAN'S MARITIME CLAIMS

TABLE OF CONTENTS

INTRODUCTION

On May 2, 1993, the Government of Iran completed legislative action on an "Act on the Marine Areas of the Islamic Republic of Iran in the Persian Gulf and the Oman Sea." On July 6, 1993, the Iran notified the Secretary General of the United Nations of the legislation.1

The legislation provides a reasonably comprehensive set of maritime claims to a territorial sea, contiguous zone, exclusive economic zone (EEZ), and continental shelf, and Iran's jurisdictional claims within those areas. Many of these claims do not comport with the requirements of international law as reflected in the 1982 United Nations Convention on the Law of the Sea (LOS Convention). The Act replaces provisions of earlier Iranian legislation:

- Act relating to the breadth of the territorial waters and to the zone of supervision,
July 19, 1934;2
- Act of April 12, 1959 amending the Act of July 15, 1934 on the territorial sea and
the contiguous zone of Iran;3
- Law of June 19, 1955 concerning the Continental Shelf;4 and

– Proclamation of October 30, 1973 concerning the outer limit of the exclusive fishing zone of Iran in the Persian Gulf and the Sea of Oman.5

1 The full text of the English translation provided by the Permanent Mission of Iran to the UN by its note

152 of July 6, 1993 is reproduced from U.N. LAW OF THE SEA BULLETIN No. 24, December 1993, at 10-15, at

Annex 1 of this study.

2 An English translation of this 1934 act may be found in UN Legislative Series, LAWS REGULATIONS ON THE

REGIME OF THE HIGH SEAS, UN Doc. ST/LEG/SER.B/1, at 81 (1951), and the French text found in UN Legislative Series, LAWS AND REGULATIONS ON THE REGIME OF THE TERRITORIAL SEA, UN Doc. ST/LEG/SER.B/6, at 24-25 (1957).

3 This 1959 act may be found in French in United Nations Legislative Series, NATIONAL LEGISLATION AND

TREATIES RELATING TO THE LAW OF THE SEA, UN Doc. ST/LEG/SER.B/15, at 88-89 (1970) and in English in

United Nations Legislative Series, NATIONAL LEGISLATION AND TREATIES RELATING TO THE LAW OF THE SEA, UN Doc. ST/LEG/SER.B/16, at 10-11 (1974).

4 The text of this 1955 law may be found in French in United Nations Legislative Series, LAWS AND

REGULATIONS ON THE REGIME OF THE TERRITORIAL SEA, UN Doc. ST/LEG/SER.B/6, at 25 (1957) and in United Nations Legislative Series, NATIONAL LEGISLATION AND TREATIES RELATING TO THE LAW OF THE SEA, UN Doc.

ST/LEG/SER.B/15, at 366-67 (1970), and in English in United Nations Legislative Series, NATIONAL

LEGISLATION AND TREATIES RELATING TO THE LAW OF THE SEA, UN Doc. ST/LEG/SER.B/16, at 151 (1974) and in

UN Office for Ocean Affairs and the Law of the Sea, LAW OF THE SEA: NATIONAL LEGISLATION ON THE CONTINENTAL SHELF, at 134 (UN Sales No. E.89.V.5, 1989).

5 The text of this 1973 proclamation may be found in English in United Nations Legislative Series,

NATIONAL LEGISLATION AND TREATIES RELATING TO THE LAW OF THE SEA, UN Doc. ST/LEG/SER.B/18, at 334-35

(1976), and in UN Office of the Special Representative of the Secretary-General for the Law of the Sea, LAW

OF THE SEA: NATIONAL LEGISLATION ON THE EXCLUSIVE ECONOMIC ZONE, THE ECONOMIC ZONE AND THE EXCLUSIVE
5

The 1993 law also references the Decree-Law No. 2/250-67 dated 31 Tir 1352 (July 22,

1973) of the Council of Ministers which established straight baselines.6

Summary of the 1993 Iran Legislation

- Most of the straight baseline segments are not drawn in accordance with the relevant principles of the international law of the sea.

- The 1993 Act makes a claim to internal water status of the water "between islands" which lacks any basis in international law and is even more expansive than earlier Iranian claims.

- It fails to limit the power to suspend the right of innocent passage as required by international law.

- Warships and certain other ships are, contrary to international law, required to receive prior approval to engage in innocent passage.

- The 1993 Act provides for supplementary laws to limit further the right of innocent passage in ways not permitted by the law of the sea.

- It claims excessive criminal and civil jurisdiction within Iran's maritime zones.

- The 1993 Act prohibits military activities within the entire EEZ, which is consistent with the freedoms of navigation and over flight enjoyed by all States in the EEZ.

- The other delimitation provisions are consistent with the law of the sea.

Analysis of Marine Areas Act, 1993

The Act consists of 23 articles in four parts: Territorial Sea, Contiguous Zone, Exclusive Economic Zone and Continental Shelf, and Final Provisions. The analysis of each article of the 1993 Marine Areas Act is preceded by the English text of that article in italics.

FISHERY ZONE AT 156, (UN Sales No. E.85.V.10, 1986).

6 The text of this 1973 Decree-Law may be found in French, with a list of coordinates, in United Nations Legislative Series, National Legislation and Treaties Relating to the Law of the Sea, UN Doc.

ST/LEG/SER.B19, at 55 (1980), and in English, including the coordinates of the basepoints and a map provided by Iran to the UN, in United Nations Office of Ocean Affairs and the Law of the Sea, THE LAW OF THE

SEA–BASELINES: NATIONAL LEGISLATION WITH ILLUS-TRATIVE MAPS, at 194-195 (Sales No. E.89, V.10, 1989)

(the longitude value for point 9 should be 55°16'55"). The English translation is reproduced in Annex 2 of this study.

6

Part I, "Territorial Sea," consists of 11 articles.

Article 1
Sovereignty

The sovereignty of the Islamic Republic of Iran extends, beyond its land territory, internal waters and its islands in the Persian Gulf, the strait of Hormuz and the Oman Sea, to a belt of sea, adjacent to the baseline, described as the territorial sea.

This sovereignty extends to the air space over the territorial sea as well as to its bed and subsoil.

The text of article 1 is drawn largely from Article 2 of the LOS Convention. The emphasis on "its islands in the Persian gulf, the strait of Hormuz and the Oman Sea" is more specific than that contained in article 5 of Iran's act of 12 April 1959, and thus would appear to be calculated to enhance its claim of sovereignty over those islands which are disputed.7

Normally, a coastal State's "islands" are considered part of its "land territory," as is reflected in Article 121 of the LOS Convention.8

Article 2
Outer Limit

The breadth of the territorial sea is 12 nautical miles, measured from the baseline. Each nautical mile is equal to 1852 meters.

The islands belonging to Iran, whether situated within or outside its territorial sea, have, in accordance with this Act, their own territorial sea.

Article 2 continues Iran's claim to a 12-mile9 territorial sea, first made in article 3 of the

1959 Act. This is the maximum breadth allowable under Article 3 of the LOS Convention.

The second paragraph of article 2 once again emphasizes the "islands belonging to

Iran."Although the LOS Convention provides special rules for the "low-tide elevations" (Art. 13) and "rocks" (Art. 121), a coastal State's islands always have their own territorial sea.

The fact that Iran chose to add this language appears to further emphasize Iran's claims

over these islands, some of which are disputed, and their marine resources.

7 Both the United Arab Emirates and Iran claim sovereignty over the islands Abu Musa (Jazireh-ye Abu Musa), Tunb al Kubra (Jazireh-ye Tonb-e Bozorg), and Tunb as Sughra (Jazireh-ye Tonb-e Kuchek).

8 As reported in FBIS-NEA-93-076 on April 20, 1993, Mr. Hasan Qashqavi, a member of the Majlis' foreign policy committee, explained that the act "also clarifies the status of the areas beyond our land territory, the internal waters, and the islands in the Persian Gulf, the Strait of Hormuz, and the Sea of Oman."

9 All miles in this study refer to nautical miles. One nautical mile equals 1,852 meters.

7

Article 3
Baseline

In the Persian Gulf and the Oman Sea, the baseline from which the breadth of the territorial sea is measured, is that one determined in the Decree No. 2 250-

67 dated 31 Tir 1352 (July 22nd, 1973) of the Council of Ministers (annexed to this Act); in other areas and islands, the low-water line along the coast constitutes the baseline. Waters on the landward side of the baseline of the territorial sea, and waters between islands belonging to Iran, where the distance of such islands does not exceed 24 nautical miles, form part of the internal waters and are under the sovereignty of the Islamic Republic of Iran.

The first subparagraph of article 3 reasserts Iran's baseline claim of July 21, 1973.10 It does not list the coordinates, but incorporates them by reference to the earlier decree.

The segments of this straight baseline system, for the most part, do not comply with international law as reflected in the requirements of Article 7 of the LOS Convention.

Rarely is the Iranian coastline "deeply indented" or fringed by islands as Article 7 requires.

Indeed, in the vicinity of most segments, the Iranian coastline is quite smooth. While the

LOS Convention does not set a maximum length, many of the segments are excessively long (13 of the 21 segments are between 25 and 114 miles long). While there is no consistent State practice on the maximum length of straight baseline segments, the United

States believes that the maximum length of an appropriately drawn straight baseline segment normally should not exceed 24 miles.

The appropriate baseline for virtually all of the Iranian coast in the Persian Gulf and the Gulf of Oman is the "normal baseline," the low-water line. The Act does use the low-water line along the west coast of Iran in the Strait of Hormuz, and for short distances along the southern coasts of Jazireh-ye11 Kish and Jazireh-ye Larak.

Baseline segment distances are given in Table 1. 10 See Annex 2 of this study for a listing of the base points.

11 "Jazireh-ye" is Persian for "island."

8

TABLE 1
Iran's Straight Baseline Segments
Group "A": Shatt al Arab (Shatt al' Arab, Arvand Rud) to Jazireh-ye Kish
Segment Length (Miles)
1-2 48 (approx.)
2-3 61.3
3-4 101.5
4-5 113.4
5-6 41.2
Group "B": Jazireh-ye Kish to Jazireh-ye Larak to the Rud-khaneh-ye Shirin
7-8 43.1
8-9 26.7
9-10 31.6
10-11 29.8
12-13 12.8
13-14 8.0
Group "C": Gulf of Oman - Damagheh-ye12 Kuh to Khalij-e13 Gavater (Gwatar Bay)
15-16 24.8
16-17 18.6
17-18 55.4
18-19 26.6
19-20 34.0
20-21 11.6

21-22 2.6
22-23 8.3
23-24 45.7
24-25 11 (approx.)

Group "A": Shatt al Arab (Shatt al 'Arab, Arvand Rud) to Jazireh-ye Kish

Segment 1-2 (see map 1) encloses the Khowr-e Musa from the intersection of the thalweg of the Shatt al Arab and a line joining the two banks of the mouth of the Shatt al Arab (Point 1) to Damagheh-ye Bahrgan (Point 2).14 The exact location of basepoint No. 1 is uncertain. However, a basepoint must be located on land. Thus, Point 1, which is situated in the entrance to the Shatt al Arab, is not in accordance with international law.15 The12 "Damagheh-ye" means "cape" in Persian.

13 "Khalij-e" means "bay" in Persian.

14 Using DMA chart NO 62434, 6th ed., Apr.21/1990, one can estimate a possible position at 29°54'50"N, 48°36'52"E.

15 It should be noted that the maritime boundary between Iran and Iraq has not been negotiated.

9

Damagheh-ye Bahrgan does not qualify as a juridical bay; a juridical bay does exist landward of this claimed straight baseline.

Segment 2-3 parallels a smooth coast which is not deeply indented. Jazireh-ye Khark, on which Point 3 is located, is not a "fringe of islands". The effect of this 61-mile long segment is to push seaward the territorial sea limit as much as 20 miles from the low-water line, encompassing an area of 868 square kilometers (258 square nautical miles) as internal waters, which should be either territorial sea or EEZ.

Segment 3-4, over 100 miles long from Jazireh-ye Khark to Jazireh-ye Nakhilu which abuts the mainland. Two small bays along the coast landward of this segment could be closed as juridical bays. This straight baseline pushes the territorial seaward an additional 12 miles along a stretch of about 35 miles, adding about 2,650 square kilometers (775 square nautical miles) to Iran's territorial sea which would otherwise be EEZ. In addition, this straight base-line carves out about 3,250 square kilometers (950 square nautical miles) of internal waters that should be either territorial sea or EEZ.

Segment 4-5, the longest segment at 113 miles, parallels a smooth coastline with a slight curvature. Given the length of this baseline segment, the effect of this claim gives to Iran approximately 3,000 square kilometers (875 square nautical miles) of territorial sea that should be high seas or EEZ, and about 4,050 square kilometers (1,180 square nautical miles) of internal waters that should be either territorial sea or EEZ.

Segment 5-6 connects Jazireh-ye Lavan and Jazireh-ye Kish, skipping Jazireh-ye

Hendorabi. Because this 41-mile long segment lies just off the coastline for most of its length, it has only a slight effect on the outer limit of the territorial sea.

A small portion of the low water line on Jazireh-ye Kish is used where the general direction of the coastline turns eastward.

Group "B": Jazireh-ye Kish to Jazireh-ye Larak to the Rudkhaneh-ye Shirin

Segment 7-8 begins at the south-east coast of Jazireh-ye Kish and extends due east to Ra's-e Shenas. This area of Iran is neither deeply indented nor fringed with islands. This 43-mile long segment has only a minimal effect on the outer limit of the territorial sea. The islands of Forur, Bani Forur and Sirri have not been included in the straight baseline system.

Segment 8-9 encloses the mouth of the Tor'eh-ye Khowran with a 27-mile line.A more appropriate line would be from Ra's-e Dastakan, on the western tip of Jazireh-ye Qeshm, northwest to the mainland in the vicinity of Ra's osh Shavari. Iran has not claimed any straight baselines to the Tumbs and Abu Musa, which are disputed with the United Arab Emirates.

10

Segments 9-10, 10-11, 12-13, and 13-14 connect islands in the northern side of the Strait of Hormuz. Segment 10-11, 30 miles long, pushes the territorial sea limit in the Strait to the

Iran-Oman continental shelf boundary, as provided for in Article 4. The low-water line is used between points 11 and on Larak Island. Segments 12-13 and 13-14 have very little effect on the seaward extent of the territorial sea in this part of the Strait of Hormuz.

The low water-line is then used for the western coastline of Iran in the eastern Strait of Hormuz until Damagheh-ye Kuh.

Group "C": Gulf of Oman - Damagheh-ye Kuh to Khalij-e Gavater (Gwatar Bay)

This group of straight baseline segments begins at the point where the Iranian coastline turns eastward to face on the Gulf of Oman (see map 2).

Segment 15-16, 25 miles long, encloses a part of the coast that is neither deeply indented nor fringed with islands. The segment pushes the territorial sea limit, at one point, about two miles further seaward than what would result using the low-water line.

Segment 16-17, 19 miles long, encloses a mere curvature in the coastline that does not qualify as a juridical bay.

Segment 17-18, 55 miles long, encloses a gently curving coastline. Straight baselines would be appropriate only for the area immediately east of point 17, where the mainland is fringed with islands.

Segments 18-19 and 19-20, 27 and 34 miles long, respectively, are situated where the coastline has only slight curvatures.

Segment 20-21, 12 miles long, encloses Khalij-e Pozm and shallow headlands forming the western tip of the mouth of Khalij-e Chah Bahar.A shorter line could be drawn to enclose Khalij-e Pozm, a juridical bay.

Segment 21-22, less than 3 miles in length, encloses a shallow concave coastline.

Segment 22-23, 8 miles long, properly encloses Khalij-e Chah Bahar, a small juridical bay.

Segment 23-24, 46 miles long, is situated where the coastline is smooth, with no fringing islands.

Segment 24-25, 11 miles long, encloses part of Khalij-e Gavater (Gwatar Bay). Point 25 is located at sea on the line joining shorelines at the mouth of the bay which Iran shares with Pakistan. Iran's maritime boundary with Pakistan has not been delimited.

The second paragraph of article 3, as it relates to islands, restates a claim first made in article 6 of the April 1959 Act that "the waters between the islands belonging to Iran 11 situated at a distance not exceeding 12 nautical miles from one another shall constitute the internal waters of Iran", while expanding the distance to 24 miles. Both claims lack any basis in international law. While waters on the landward side of baselines are internal, international law makes no provision for making "waters between islands . . .internal waters." Islands have their own territorial sea, but, except for situations where the islands are part of a valid straight baseline system or closing line for a juridical bay, they do not define internal waters.

The 1993 Act drops the unusual claim made in article 5 of the 1959 Act that "islands situated at a distance not exceeding 12 nautical miles from one another shall be considered as a single island and the limit of their territorial sea shall be determined from the islands remotest from the center of the archipelago."Article 3 of the 1934 law had similarly claimed that "the islands comprising an archipelago shall be deemed to form a single island and the breadth of the territorial waters shall be measured from the islands remotest from the center of the archipelago."

Article 4
Delimitation

Wherever the territorial sea of Iran overlaps the territorial seas of the states with opposite or adjacent coasts, the dividing line between the territorial seas of Iran and those states shall be, unless otherwise agreed between the two parties, the median line every point of which is equidistant from the nearest point on the baseline of both states.

The language of article 4 replaces article 4 of the 1959 Act, and is unobjectionable. While article 4 reiterates only the first sentence of Article 15 of the LOS Convention, and therefore does not address the "special circumstances" exceptions to the equidistance rule, the inclusion of Article 15's language, "unless otherwise agreed between the two parties," mitigates this concern. Moreover, Iran has been, in practice, willing to negotiate appropriate continental shelf boundaries with its neighbors. For example, Iran has negotiated continental shelf boundaries, which the States involved may or may not recognize as coterminous with territorial sea boundaries, with Saudi Arabia, Qatar, Bahrain, United Arab Emirates (Dubai), and Oman. See the discussion of Article 19 below.

Article 5
Innocent Passage

The passage of foreign vessels, except as provided for in Article 9, is subject to the principle of innocent passage so long as it is not prejudicial to good order, peace and security of the Islamic Republic of Iran. Passage, except as in cases of force majeure, *shall be continuous and expeditious.*
12

Article 5 is new. Earlier Iranian laws contained no provisions concerning innocent passage. The new text is largely unobjectionable. While not tracking the specific text of Articles 19(1) and 19(2) of the LOS Convention, this article uses all the key terms. The one problem is the reference to article 9 which, as discussed below, contains objectionable constraints on the right of innocent passage of warships and certain other vessels.

Article 6
Requirements of Innocent Passage

Passage of foreign vessels, in cases when they are engaged in any of the following activities shall not be considered innocent and shall be subject to relevant civil and criminal laws and regulations:

a) Any threat or use of force against the sovereignty, territorial integrity or political independence of the Islamic Republic of Iran, or in any other manner in violation of the principles of international law;

b) Any exercise or practice with weapons of any kind;

c) Any act aimed at collecting information prejudicial to national security, defense or economic interests of the Islamic Republic of Iran;

d) Any act of propaganda aimed at affecting the national security, defense or economic interests of the Islamic Republic of Iran;

e) the launching, landing or transferring on board of any aircraft or helicopter, or any military devices or personnel to other vessel or to the coast;

f) the loading or unloading of any commodity, currency or person contrary to the laws and regulations of the Islamic Republic of Iran;

g) Any act of pollution of the marine environment contrary to the rules and regulations of the Islamic Republic of Iran;

h) Any act of fishing or exploitation of the marine resources;

i) the carrying out of any scientific research and cartographic and seismic Surveys or sampling activities;

j) Interfering with any systems of communication or any other facilities or installations of the Islamic Republic of Iran;

k) Any other activity not having a direct bearing on passage.

Article 6 contains quite a few variations from the list of activities in Article 19 of the LOS

Convention that make passage not innocent, several of which are objectionable.

While Article 19(2)(c) of the LOS Convention lists "any act aimed at collecting information to the prejudice of the defense or security of the coastal State," subparagraph c) of the 1993 Iranian Act uses the term "national security" and adds "economic interests". These additions do not seem to be at variance with the purpose and intent of Article 19(2)(c).

13

While Article 19(2)(d) lists "any act of propaganda aimed at affecting the defense or security of the coastal State," subparagraph d) of the Act again uses the term "national security" and adds "economic interests". These additions do not seem to be at variance with the purpose and intent of Article 19(2)(d).

Article 19(2)(e) lists "the launching, landing or taking on board of any aircraft", and Article 19(2)(f) proscribes similar acts with regard to "any military device". Subparagraph e) of the Act adds the transfer of "personnel" to other vessels or to the coast, an act not mentioned in Article 19(2)(e) and (f) of the LOS Convention, although arguably consistent with the "catch-all" provision of Article 19(2)(l).

Article 19(2)(g) refers to "the loading or unloading of any commodity, currency or person contrary to the customs, fiscal, immigration or sanitary laws and regulations of the coastal State". Subparagraph f) of the Act omits the limiting words "customs, fiscal, immigration or sanitary" and thus would treat the loading or unloading of commodities, etc., contrary to any Iranian law, as making passage not innocent.

While Article 19(2)(h) of the LOS Convention only makes "willful and serious pollution" non-innocent actions, subparagraph g) of the Iranian Act is more expansive, proscribing "any act of pollution of the marine environment . . ."

The addition in subparagraph h) of the Act of exploitation of marine resources to fishing (Article 19(2)(i) of the LOS Convention) is unobjectionable since such acts have no direct bearing on passage and clearly fall within the sovereignty of the coastal state.

Subparagraph i) of the Act limits "research or survey activities" (Article 19(2)(j) of the LOS Convention) to cartographic and seismic surveys and sampling activities.

Article 7
Supplementary Laws and Regulations
The Government of the Islamic Republic of Iran shall adopt such other regulations as are necessary for the protection of its national interests and the proper conduct of innocent passage.

Article 21 of the LOS Convention permits the coastal State to adopt laws and regulations, in conformity with the provisions of the LOS Convention and other rules of international law, relating to innocent passage, in respect to eight categories.Article 7 of the 1993 Iranian Act appears to be too broadly cast. The language "necessary for the protection of its national interests and the proper conduct of innocent passage" does not contain the caveats of Articles 21(2) and 24 of the LOS Convention which prohibit the application of such laws to design, construction, manning or equipment of foreign ships unless they are giving effect tom generally accepted international rules or standards. The LOS Convention also prohibits the

imposition of requirements that have the practical effect of denying or impairing the right of

14 innocent passage or of discriminating in form or in fact against the ships of any State or against ships carrying cargoes to, from or on behalf of any State.

15

Article 8
Suspension of Innocent Passage

The Government of the Islamic Republic of Iran inspired by its high national interests and to defend its security may suspend the innocent passage in parts of its territorial sea. Article 8 is objectionable in that it fails to state the two limitations on the right of a coastal State to suspend innocent passage. First, any such suspension must be temporary; and second, Iran must first publish the details of the suspension (time and place). It would have been preferable for Iran to have simply adopted the very comprehensive language of Article 25(3) of the LOS Convention.

Article 9
Exceptions to Innocent Passage

Passage of warships, submarines, nuclear-powered ships and vessels or any other floating objects or vessels carrying nuclear or other dangerous or noxious substances harmful to the environment, through the territorial sea is subject to the prior authorization of the relevant authorities of the Islamic

Republic of Iran. Submarines are required to navigate on the surface and to show their flag.

The first sentence of article 9 is objectionable. The LOS Convention does not permit a coastal State to require a foreign vessel to seek the prior authorization of, or notification to, the coastal State as a condition of conducting innocent passage through its territorial sea. Iran's signature of the LOS Convention was also accompanied by an objectionable declaration which included its claim to require "prior authorization for warships willing to exercise the right of innocent passage through the territorial sea".16 In the United States' view, such a "declaration" is tantamount to a reservation; however, Article 309 of the Convention specifically prohibits reservations. The United States has previously protested this claim17 and on many occasions since 1989 U.S. warships have exercised the right of innocent passage through the Iranian territorial sea without notice to or reaction from Iran.18

16 Iran's declaration stated: "In the light of customary international law, the provisions of article 21, read in association with article 19 (on the Meaning of Innocent Passage) and article 25 (on the Rights of Protection

of the Coastal States), recognizes (though implicitly) the rights of the Coastal States to take measures to safeguard their security interests including the adoption of laws and regulations regarding, *inter alia,* the requirements of prior authorization for warships willing to exercise the right of innocent passage through the territorial sea." UN, MULTILATERAL TREATIES DEPOSITED WITH THE SECRETARY-GENERAL: STATUS AS AT 31

DECEMBER 1992, UN Doc. ST/LEG/SER.E/11, at 769 (1993).

17 See Limits in the Seas No. 112, at 59.

18 See Secretary of Defense, ANNUAL REPORT TO THE PRESIDENT AND THE CONGRESS 78 (1992); *id.* at 85 (1993) and *id.,* Appendix G (1994) for assertions conducted by DoD assets between October 1, 1990 and September 30, 1991; between

October 1, 1991 and September 30, 1992; and between October 1, 1992, and September 30, 1993, respectively.

16

Further, while the LOS Convention discusses nuclear-powered ships and vessels carrying dangerous cargoes, Article 23 simply requires that they carry appropriate documents and observe special precautionary measures established for such ships by international agreements. The second sentence of article 9 is unobjectionable; it simply reiterates the relevant language of Article 20 of the LOS Convention that submarines must navigate on the surface while engaged in innocent passage.19

Article 10
Criminal Jurisdiction

In the following cases, the investigation, prosecution and punishment in connection with any crimes committed on board the ships passing through the territorial sea is within the jurisdiction of the judicial authorities of the Islamic Republic of Iran:

a) if the consequences of the crime extend to the Islamic Republic of Iran;

b) If the crime is of a kind to disturb the peace and order of the country or the public order of the territorial sea;

c) if the master of the ship or a diplomatic agent or consular officer of the flag state asks for the assistance and investigation;

d) if such investigation and prosecution is essential for the suppression of illicit traffic in narcotic drugs or psychotropic substances.

Depending on how it is interpreted and applied, article 10 could be objectionable.

Superficially, it simply tracks the language of Article 27(1) of the LOS Convention.

However, the text raises concerns. Article 27(1) of the LOS Convention starts with the presumption that the "criminal jurisdiction of the coastal State should not be exercised on board a foreign ship passing through the territorial sea . . .save only in the following cases .

. ." The Iranian statute starts with the idea that "the investigation, prosecution, and punishment in connection with any crimes . . .is within the jurisdiction of the judicial authorities of the Islamic Republic of Iran . . ." While unlikely, Iranian courts could apply liberally the first two sets of cases, "consequences of the crime" and "of a kind to disturb the peace and order of the country . . ." to claim broad criminal jurisdiction over persons on board ships passing through Iranian waters.

Second, Article 27 of the LOS Convention is located in a subsection dealing with "merchant ships and government ships operated for commercial purposes." In contrast, the Iranian Act does not limit its scope. Although it would be inconsistent with the international law principle of sovereign immunity to do so, under the Act Iran could claim.

19 The United States and Russia have expressed their views on innocent passage in a 1989 joint statement reproduced as Annex III to Limits in the Seas No. 112.

17 right to investigate, prosecute, and punish "any crimes" even on board warships and other government vessels.

Article 11
Civil Jurisdiction

The competent authorities of the Islamic Republic of Iran may stop, divert or detain a ship and its crew for the enforcement of attachment orders or court judgments if:

a) the ship is passing through the territorial sea after leaving the internal waters of Iran;

b) the ship is lying in the territorial sea of Iran;

c) the ship is passing through the territorial sea, provided that the origin of the attachment order or court judgment rests in the obligations or requirements arising from the civil liability of the ship itself.

Article 11 is also of concern. Article 11(b) permits Iran to "stop, divert, or detain a ship and its crew for the enforcement of attachment orders or court judgments if .the ship is lying in the territorial sea of Iran . . .." This is much too broad a claim to civil jurisdiction. Article 28(2) of the LOS Convention provides that "[t]he coastal State may not levy execution against or arrest the ship for the purpose of any civil proceedings, save only in respect of obligations or liabilities assumed or incurred by the ship itself in the course or for the purpose of its voyage through the waters of the coastal State."

Article 11 also is defective in that it does not purport to limit its scope to commercial vessels. Any attempt to assert civil jurisdiction over or to detain a sovereign immune vessel would be a clear violation of Article 32 of the LOS Convention.

Part II, "Contiguous Zone," consists of two articles, 12 and 13.

Article 12
Definition

The contiguous zone is an area adjacent to the territorial sea the outer limit of which is 24 nautical miles from the baseline.

In 1934 Iran claimed the functional equivalent of a 12-mile contiguous zone, calling it a "zone of marine supervision."Article 12 of the 1993 Act is unobjectionable. Following the word "baseline(s)" the text would have been clearer had Iran added the language from Article 33(2) of the LOS Convention "from which the breadth of the territorial sea is measured."

18

Article 13
Civil and Criminal Jurisdiction

The Government of the Islamic Republic of Iran may adopt measures necessary to prevent the infringement of laws and regulations in the contiguous zone, including security, customs, maritime, fiscal, immigration, sanitary and environmental laws and regulations and investigation and punishment of offenders.

Article 13 is objectionable. International law as reflected in Article 24 of the 1958 Geneva Convention on the Territorial Sea and the Contiguous Zone and Article 33(1) of the LOS Convention permits a coastal State to exercise the control necessary to prevent

or punish infringement of only four categories of offenses: violations of customs, fiscal, immigration, and sanitary (health and quarantine) laws and regulations. Article 13 is inconsistent with international law by adding the additional categories of "security," "maritime," and "environment." Indeed, article 13 seems designed to establish a "security zone". International law does not recognize the right of a coastal State to establish a security zone in peacetime that would restrict the exercise of free navigation and over flight beyond the territorial sea. Nevertheless, Iran's 1934 law claimed the zone of supervision existed "with a view to ensuring the operation of certain laws and conventions concerning the security and protection of the country and its interests or the safety of navigation." Iran consistently sought (and failed) during the Third UN LOS Conference to include "security" as an interest to be protected in the contiguous zone, and continued that claim in its declaration accompanying its signature of the LOS Convention. On the other hand, the claim to be able to prevent the violation of "maritime . . .laws and regulations" in the contiguous zone is new, and raises concerns because of its potentially expansive application. Does this include cases which arise in admiralty? Does it permit Iran to restrict maritime navigation? Finally, adding environment jurisdiction in the contiguous zone is both new and cause for concern. While the coastal State has certain authority to restrict vessel-source pollution within its territorial sea and EEZ, direct enforcement requires that the pollution have occurred in the territorial sea. If the vessel pollutes beyond the territorial sea, the coastal State may only require that the vessel provide information, unless the discharge causes or threatens major damage. See LOS Convention, Article 220.
19

Part III, "Exclusive Economic Zone and Continental Shelf," is comprised of eight articles.

Iran first claimed its continental shelf in 1955, and reiterated that claim in its 1973 proclamation. The 1973 proclamation also claimed an exclusive fishery zone coextensive with the outer limits of its continental shelf, or 50 miles in the Sea of Oman.

Article 14
Sovereign Rights and Jurisdiction in the Exclusive Economic Zone

Beyond its territorial sea which is called the exclusive economic zone, the Islamic Republic of Iran exercises its sovereign rights and jurisdiction with regard to:

a) exploration, exploitation, conservation and management of all natural resources, whether living or non-living, of the sea-bed and subsoil thereof and its superjacent waters, and with regard to other economic activities for the production of energy from water, currents and winds.

b) adoption and enforcement of appropriate laws and regulations especially for the following activities:

1) the establishment and use of artificial islands and other installations and structures, laying of submarine cables and pipelines and the establishment of relevant security and safety zones;

2) any kind of research;

3) the protection and preservation of the marine environment.

c) such sovereign rights as granted by regional or international treaties.

By Article 14, Iran became the 88th state to claim an EEZ. While article 14 largely comports with Article 56 of the LOS Convention, the United States does not accept Iran's declaration, filed upon signing the Convention, that the "notion of the 'exclusive economic zone'" was new and available only to States Party to the Convention. The ICJ has consistently ruled, since the Gulf of Maine delimitation case in 1984, that the exclusive economic zone is established customary law.20

The text of article 14 raises three concerns.

First, article 14(b)(1) could be read to impermissibly claim the right to establish within the EEZ "relevant security and safety zones." While Article 60 of the LOS Convention

20 *Case Concerning Delimitation of the Maritime Boundary of the Gulf of Maine (Canada/United States),*

[1984] I.C.J. Rep. 246, 294, at para. 94; *Case Concerning the Continental Shelf (Tunisia/Libya)*, [1982] ICJ

Rep. 74, at para. 100; *Case Concerning the Continental Shelf (Libya/ Malta)*, [1985] ICJ Rep. 33, at para. 34;

Case Concerning the Maritime Delimitation in the Area between Greenland and Jan Mayen, judgment para.

47-48 (June 14, 1993).Accord, B. Kwiatkowska, THE 200 MILE EXCLUSIVE ECONOMIC ZONE IN THE NEW LAW OF THE SEA 27-37 (1989) and D. Attard, THE EXCLUSIVE ECONOMIC ZONE IN INTERNATIONAL LAW 277-309 (1987).

20 contemplates small safety zones around artificial islands, installations and structures located in the EEZ, no provision of the LOS Convention permits security zones in the EEZ.

Article 58(1) of the LOS Convention sets forth the rule that there exist within the EEZ the high seas freedoms of navigation and over flight and other internationally lawful uses of the seas.

Second, in article 14(b)(1), Iran impermissibly claims the right to control the laying of submarine cables and pipelines in its EEZ. Article 58(1) of the LOS Convention sets forth the rule that there exists within the EEZ the high seas freedoms of laying submarine cables and pipelines.21 Third, in article 14(b)(2), Iran claims the right to control "any kind of research."Article 56(1)(b)(ii) of the LOS Convention gives the coastal State jurisdiction with regard to "marine scientific research" (MSR). The term MSR is not defined in the LOS Convention. The United States accepts that MSR is the general term most often used to describe those activities undertaken in the ocean and coastal waters to expand scientific knowledge of the marine environment. MSR includes oceanography, marine biology, fisheries research, scientific ocean drilling, geological/geophysical scientific surveying, as well as other activities with a scientific purpose. It may be noted, however, that "survey activities," "prospecting" and "exploration" are primarily dealt with in other parts of the LOS Convention, notably Parts II, III, XI and Annex III, rather than Part XIII. "This would indicate that those activities do not fall under the regime of Part XIII."22 When activities similar to those mentioned above as MSR are conducted for commercial resource purposes, most governments, including the United States, do not treat them as MSR. Additionally, activities such as hydrographic surveys,23 the purpose of which is to obtain information for the making of navigational charts, and the collection of information that, whether or not classified, is to be used for military purposes, are not considered by the United States to

21 On the other hand, Article 79(3) of the LOS Convention makes the "delineation of the course for the laying of submarine pipelines on the continental shelf" subject to the consent of the coastal State.

22 UN Office for Oceans Affairs and the Law of the Sea, LAW OF THE SEA: MARINE SCIENTIFIC RESEARCH: A GUIDE TO THE IMPLEMENTATION OF THE RELEVANT PROVISIONS OF THE UNITED NATIONS CONVENTION ON THE LAW OF THE SEA 1 para. 2 (UN Sales No. E.91.V.3, 1991). See also, id., LAW OF THE SEA: NATIONAL LEGISLATION, REGULATIONS AND SUPPLEMENTARY DOCUMENTS ON MARINE SCIENTIFIC RESEARCH IN AREAS UNDER NATIONAL JURISDICTION (UN Sales No. E.89.V.9, 1989). The United States does not claim jurisdiction over MSR in its EEZ.

23 The International Hydrographic Organization has defined hydrographic surveys as "the science of measuring and depicting those parameters necessary to describe the precise nature and configuration of the sea-bed and coastal strip, its geographical relationship to the land-mass, and the characteristics and dynamics of the sea." Definition 40, Glossary of Technical Terms Used in the United Nations Convention on the Law of the Sea, Appendix I of UN Office for Ocean Affairs and the Law of the Sea, THE LAW OF THE SEA: BASELINES: AN EXAMINATION OF THE RELEVANT PROVISIONS OF THE UNITED NATIONS CONVENTION ON THE LAW OF THE SEA 56 (UN Sales No. E.88.V.5*, 1989).

21

be MSR and, therefore, are not subject to coastal state jurisdiction.24 To the extent that Iran interprets "any kind of research" beyond MSR as defined in the Convention, they would interfere with the high seas freedoms of other States, including the right to conduct hydrographic surveys.

Article 15
Sovereign Rights and Jurisdiction in the Continental Shelf

The provisions of Article 14 shall apply mutatis mutandis *to the sovereign rights and jurisdiction of the Islamic Republic of Iran in its continental shelf, which comprises the sea-bed and subsoil of the marine areas that extend beyond the territorial sea throughout the natural prolongation of the land territory.*

Article 15 purports to apply the same rules in the EEZ to the continental shelf. Since Iran is shelf- and EEZ-locked (because boundaries required with neighboring States prevent Iran from claiming a maximum breadth of EEZ or continental shelf), this makes some sense.

However, the same objections discussed above concerning security, safety, and research apply to the continental shelf as well.25

Article 16
Prohibited Activities

Foreign military activities and practices, collection of information and any other activity inconsistent with the rights and interests of the Islamic Republic of Iran in the exclusive economic zone and the continental shelf are prohibited.

Article 16 is perhaps the most objectionable article in the Act. It purports to establish a security zone for the entire EEZ, restricting freedom of military navigation, over flight, and other activities. Its blanket prohibition of "foreign military activities and practices . . . inconsistent with the rights and interests" of Iran within the EEZ is clearly inconsistent with the LOS Convention. Article 58 of the LOS Convention affirms that all States have the high seas freedoms listed in Article 87 of the Convention. A coastal State does not have the right to prohibit or restrict foreign military activ-

ities within the EEZ during times of peace. 24 1989 State telegram 122770. In Part XII of the Convention regarding protection and preservation of the marine environment, Article 236 provides that the environmental provisions of the Convention do not apply to warships, naval auxiliaries, and other vessels and aircraft owned or operated by a nation and used, for the time being, only on government non-commercial service. The provisions of Part XIII regarding marine scientific research similarly do not apply to military activities. Oxman, *The Regime of Warships Under the United Nations Convention on the Law of the Sea,* 24 VA. J. INT'L L. 809, 844-47 (1984). See also Negroponte, *Current Developments in U.S. Oceans Policy,* Dep't. St. Bull., Sep. 1986, at 86. U.S. policy is to encourage freedom of MSR. 25 With regard to submarine cables and pipelines, Article 79 of the LOS Convention permits limited coastal state jurisdiction.
22

Article 17
Scientific Activities, Exploration and Research

Any activity to recover drowned26 objects and scientific research and exploration in the exclusive economic zone and the continental shelf is subject to the permission of the relevant authorities of the Islamic Republic of Iran.

Article 17 is unnecessary. While the coastal State has the right to regulate, authorize, and conduct MSR in its EEZ and on its continental shelf (within reasonable limits), the reference to "recover drowned objects" is potentially too broad. If it refers to meteorological gear, sonobuoys, torpedoes, or other equipment used in weapons tests, or other items used in military exercises, it is objectionable; these activities are not subject to coastal State regulation. If it is in reference to MSR or equipment used in resource exploitation (e.g., core samplers, seismic test gear, fishing nets and traps), it comports with international law.

Article 18
Preservation of Environment and Natural Resources

The Government of the Islamic Republic of Iran shall take appropriate measures for the protection and preservation of the marine environment and proper exploitation of living and other resources of the exclusive economic zone and the continental shelf.

Article 18 is unobjectionable.

Article 19
Delimitation

The limits of the exclusive economic zone and the continental shelf of the Islamic Republic of Iran, unless otherwise determined with bilateral agreements, shall be a line every point of which is equidistant from the nearest point on the baselines of two states.

The language in article 19, "unless otherwise agreed between the two parties," takes into account the fact that Iran has negotiated continental shelf boundaries with Saudi Arabia, Qatar, Bahrain, United Arab Emirates (Dubai), and Oman.27 Iran does not have boundary agreements with other parts of the Emirates, Iraq, Kuwait or Pakistan.

26 Probably "sunken" objects.

27 See *Limits in the Seas* Nos. 24, 25, 58, 63, 67 and 94 (1970, 1974, 1976 and 1981); 1 INTERNATIONAL

MARITIME BOUNDARIES 315-20 (Charney and Alexander eds. 1993); and 2 *id.* 1481 *et seq.*

23

Article 19 is unobjectionable in practical result, if not fully consistent with the LOS Convention. To the extent it refers to bilateral agreements, it is consistent with Articles 74 and 83 of the LOS Convention.

Article 20
Civil and Criminal Jurisdiction

The Islamic Republic of Iran shall exercise its criminal and civil jurisdiction against offenders of the laws and regulations in the exclusive eco-

nomic zone and continental shelf and shall, as appropriate, investigate or detain them.

Article 20 is unobjectionable, assuming it is applied consistent with Article 73 of the LOS Convention.

Article 21
Right of Hot Pursuit

The Government of the Islamic Republic of Iran reserves its right of hot pursuit against offenders of laws and regulations relating to its internal waters, territorial sea, contiguous zone, exclusive economic zone and the continental shelf, in such areas and the high seas.

Article 21 is unobjectionable, assuming it is applied consistent with Article 111 of the LOS Convention.

Part IV, "Final Provisions," is composed of two articles.

Article 22
Executive Regulations

The Council of Ministers shall specify the mandates and responsibilities [powers and duties] of different ministries and organizations charged with the enforcement of this Act. The said ministries and organizations shall, within one year after the approval of this Act, prepare the necessary regulations and have them approved by the Council of Ministers.

Pending the adoption of new executive regulations, the existing rules and regulations shall remain in force.

Article 22 is unobjectionable. The goal is for the implementing regulations to be completed within one year of the date of approval of the Act, May 2, 1994. Of course, Iran's "existing rules and regulations" are also objectionable in many respects.

Article 23
24

All laws and regulations contrary to the present Act, upon its ratification, are hereby abrogated.

Article 23, untitled, is unobjectionable.

Transit Passage

It is worth noting that the Act does not address passage through straits used for international navigation at all. Although Iran's declaration accompanying its signature of the LOS Convention claimed the right of transit passage through straits used for international navigation – and the "notion of 'exclusive economic zone'" – are contractual rights limited to States Party to the LOS Convention,28 it is generally agreed that the transit passage is a right of all States under international law. The Strait of Hormuz is one of the most important in the world. While Iran is not obliged to affirm the right in its national law, failure to address the right of surface ships, submarines, and aircraft to transit the strait and its approaches raises the possibility of diplomatic and operational disagreement. The commercial and naval forces of the United States and other maritime nations regularly exercise the right of transit passage through the Strait of Hormuz. The United States has previously made its views known to Iran that the right of transit passage applies in the Strait of Hormuz to the ships and aircraft of all States.29

Military Operations

Iran's excessive maritime claims cannot stand in the face of the day-to-day military operations the United States and other States are conducting in the region consistent with international law. U.S. aircraft carriers, aircraft, submarines and surface combatants, as well as other States' military units, operate regularly in the international waters of the northern and central Persian Gulf while enforcing UN sanctions against Iraq. When there was a mission to sweep mines in the Gulf, minesweeping ships and helicopters actively, sought out and destroyed mines over many months following the two Gulf wars. This was a "foreign military . . .practice" clearly inconsistent with article 16 of the Iranian Act, conducted without objection from Iran. Moreover, warships representing a wide variety of nations pass through Iran's territorial sea in innocent passage without objection from Iran, despite Iran's requirement that prior authorization be obtained for each transit. These examples of State practice, shared in by many nations and fully consistent with international law, clearly outweigh Iran's claims to restrict freedom of navigation.

Diplomatic Protest
28 UN, MULTILATERAL TREATIES DEPOSITED WITH THE SECRETARY-GENERAL: STATUS AS AT 31 DECEMBER 1992,
UN Doc. ST/LEG/SER.E/11 at 769, (1993).
29 See Limits in the Seas No. 112, at 65-67.
25
The views of the United States on Iran's 1993 Marine Areas Act were expressed in a note
to the UN Secretary General dated January 11, 1994.30
30 This note is reproduced in Annex 3 of this study.
26

ANNEX 1
Act on the Marine Areas of the Islamic Republic of Iran in the Persian Gulf and the Oman Sea31
PART I
Territorial sea

Article 1
Sovereignty
The sovereignty of the Islamic Republic of Iran extends, beyond its land territory, internal waters and its islands in the Persian Gulf, the Strait of Hormuz and the Oman Sea, to a belt of sea, adjacent to the baseline, described as the territorial sea.

This sovereignty extends to the airspace over the territorial sea as well as to its bed and subsoil.

Article 2
Outer limit
The breadth of the territorial sea is 12 nautical miles, measured from the baseline. Each nautical mile is equal to 1,352 meters.

The islands belonging to Iran, whether situated within or outside its territorial sea, have, in accordance with this Act, their own territorial sea.

Article 3
Baseline
In the Persian Gulf and the Oman Sea, the baseline from which the breadth of the territorial sea is measured is that one determined in Decree No. 2/250-67 dated 31 Tir 1352 (22

July 1973) of the Council of Ministers (annexed to this Act);2 in other areas and islands, the low-water line along the coast constitutes the baseline.

Waters on the landward side of the baseline of the territorial sea, and waters between islands belonging to Iran, where the distance of such islands does not exceed 24 nautical miles, form part of the internal waters and are under the sovereignty of the Islamic Republic of Iran.

31 Text transmitted by the Permanent Mission of the Islamic Republic of Iran to the United Nations in a note

Verbal No. 152, dated July 6, 1993. Annex 1 is reproduced here from UN Law of the Sea Bulletin No. 24, at

10-15 (Dec. 1993) (the length of a nautical mile correctly reads "1,852" in the text transmitted to the UN).

27

ANNEX 1 (cont'd)
Act on the Marine Areas of the Islamic Republic of Iran in the Persian Gulf and the Oman Sea

Article 4
Delimitation
Wherever the territorial sea of Iran overlaps the territorial seas of the States with opposite or adjacent coasts, the dividing line between the territorial seas of Iran and those states shall be, unless otherwise

agreed between the two parties, the median line every point of which is equidistant from the nearest point on the baseline of both States.

Article 5
Innocent passage
The passage of foreign vessels, except as provided for in article 9, is subject to the principle of innocent passage so long as it is not prejudicial to good order, peace and security of the Islamic Republic of Iran.

Passage, except as in cases of force majeure, shall be continuous and expeditious.

Article 6
Requirements of innocent passage
Passage of foreign vessels, in cases when they are engaged in any of the following activities, shall not be considered innocent and shall be subject to relevant civil and criminal laws and regulations:

(a) Any threat or use of force against the sovereignty, territorial integrity or political independence of the Islamic Republic of Iran, or in any other manner in violation of the principles of international law;

(b) Any exercise or practice with weapons of any kind;

(c) Any act aimed at collecting information prejudicial to the national security, defense or economic interests of the Islamic Republic of Iran;

(d) Any act of propaganda aimed at affecting the national security, defense or economic interests of the Islamic Republic of Iran;

(e) The launching, landing or transferring on board of any aircraft or helicopter, or any military devices or personnel to another vessel or to the coast;

(f) The loading or unloading of any commodity, currency or person contrary to the laws and regulations of the Islamic Republic of Iran;

28

ANNEX 1 (cont'd)

Act on the Marine Areas of the Islamic Republic of Iran in the Persian Gulf and the Oman Sea

(g) Any act of pollution of the marine environment contrary to the rules and regulations of the Islamic Republic of Iran;

(h) Any act of fishing or exploitation of the marine resources;

(i) The carrying out of any scientific research and cartographic and seismic surveys or sampling activities;

(j) Interfering with any systems of communication or any other facilities or installations of the Islamic Republic of Iran;

(k) Any other activity not having a direct bearing on passage.

Article 7
Supplementary laws and regulations

The Government of the Islamic Republic of Iran shall adopt such other regulations as are necessary for the protection of its national interests and the proper conduct of innocent passage.

Article 8
Suspension of innocent passage

The Government of the Islamic Republic of Iran, inspired by its high national interests and to defend its security, may suspend the innocent passage in parts of its territorial sea.

Article 9
Exceptions to innocent passage

Passage of warships, submarines, nuclear-powered ships and vessels or any other floating objects or vessels carrying nuclear or other dangerous or noxious substances harmful to the environment, through the territorial sea is subject to the prior authorization of the relevant authorities of the Islamic Republic of Iran. Submarines are required to navigate on the surface and to show their flag.

Article 10
Criminal jurisdiction
29

ANNEX 1 (cont'd)

Act on the Marine Areas of the Islamic Republic of Iran

in the Persian Gulf and the Oman Sea In the following cases, the investigation, prosecution, and punishment in connection with any crimes committed on board the ships passing through the territorial sea is within the jurisdiction of the judicial authorities of the Islamic Republic of Iran:

(a) If the consequences of the crime extend to the Islamic Republic of Iran;

(b) If the crime is of a kind to disturb the peace and order of the country or the public order of the territorial sea;

(c) If the master of the ship or a diplomatic agent or consular officer of the flag State asks for the assistance and investigation;

(d) If such investigation and prosecution is essential for the suppression of illicit traffic in narcotic drugs or psychotropic substances.

Article 11

Civil jurisdiction

The competent authorities of the Islamic Republic of Iran may stop, divert or detain a ship and its crew for the enforcement of attachment orders or court judgments if:

(a) The ship is passing through the territorial sea after leaving the internal waters of Iran;

(b) The ship is lying in the territorial sea of Iran;

(c) The ship is passing through the territorial sea, provided that the origin of the attachment order or court judgment rests in the obligations or requirements arising from the civil liability of the ship itself.

PART II

Contiguous zone

Article 12

Definition

The contiguous zone is an area adjacent to the territorial sea the outer limit of which is 24 nautical miles from the baseline.

Article 13
Civil and criminal jurisdiction
30

ANNEX 1 (cont'd)
Act on the Marine Areas of the Islamic Republic of Iran
in the Persian Gulf and the Oman Sea
The Government of the Islamic Republic of Iran may adopt
measures necessary to prevent the infringement of laws and regulations in the contiguous zone, including security, customs, maritime, fiscal, immigration, sanitary and environmental laws and
regulations and investigation and punishment of offenders.
31

ANNEX 1 (cont'd)
Act on the Marine Areas of the Islamic Republic of Iran
in the Persian Gulf and the Oman Sea
PART III
Exclusive economic zone and continental shelf

Article 14
Sovereign rights and jurisdiction in the exclusive economic zone
Beyond its territorial sea, which is called the exclusive economic zone, the Islamic Republic of Iran exercises its sovereign
rights and jurisdiction with regard to:
(a) Exploration, exploitation, conservation and management
of all natural resources, whether living or non-living, of the seabed
and subsoil thereof and its superjacent waters, and with regard to
other economic activities for the production of energy from water,
currents and winds. These rights are exclusive;
(b) Adoption and enforcement of appropriate laws and regulations, especially for the following activities:
(i) The establishment and use of artificial islands and other
installations and structures, laying of submarine cables and pipelines and the establishment of relevant security and safety zones;

(ii) Any kind of research;

(iii) The protection and preservation of the marine environment;

(c) Such sovereign rights as granted by regional or international treaties.

Article 15
Sovereign rights and jurisdiction in the continental shelf

The provisions of article 14 shall apply mutatis mutandis to the sovereign rights and jurisdiction of the Islamic Republic of Iran in its continental shelf, which comprises the seabed and subsoil of the marine areas that extend beyond the territorial sea throughout the natural prolongation of the land territory.

Article 16
Prohibited activities
32
ANNEX 1 (cont'd)

Act on the Marine Areas of the Islamic Republic of Iran in the Persian Gulf and the Oman Sea

Foreign military activities and practices, collection of information and any other activity inconsistent with the rights and interests of the Islamic Republic of Iran in the exclusive economic zone and the continental shelf are prohibited.

Article 17
Scientific activities, exploration and research

Any activity to recover drowned objects and scientific research and exploration in the exclusive economic zone and the continental shelf is subject to the permission of the relevant authorities of the Islamic Republic of Iran.

Article 18
Preservation of the environment and natural resources

The Government of the Islamic Republic of Iran shall take appropriate measures for the protection and preservation of the marine environment and proper exploitation of living and other resources of the exclusive economic zone and the continental shelf.

Article 19
Delimitation
The limits of the exclusive economic zone and the continental shelf of the Islamic Republic of Iran, unless otherwise determined in accordance with bilateral agreements, shall be a line every point of which is equidistant from the nearest point on the baselines of two States.

Article 20
Civil and criminal jurisdiction
The Islamic Republic of Iran shall exercise its criminal and civil jurisdiction against offenders of the laws and regulations in the exclusive economic zone and continental shelf and shall, as appropriate, investigate or detain them.

Article 21
Right of hot pursuit
The Government of the Islamic Republic of Iran reserves its right of hot pursuit against
offenders of laws and regulations relating to its internal waters, territorial sea, contiguous zone, exclusive economic zone and the continental shelf, in such areas and the high seas.
33

ANNEX 1 (cont'd)
Act on the Marine Areas of the Islamic Republic of Iran
in the Persian Gulf and the Oman Sea
PART IV
Final provisions

Article 22
Executive regulations
The Council of Ministers shall specify the mandates and responsibilities [powers and duties] of different ministries and organizations charged with the enforcement of this Act.

The said ministries and organizations shall, within one year after the approval of this Act, prepare the necessary regulations and have them approved by the Council of Ministers.

Pending the adoption of new executive regulations, the existing rules and regulations shall remain in force.

Article 23
All laws and regulations contrary to the present Act, upon its ratification, are hereby abrogated.

The above Act, comprising 23 articles, was ratified at the plenary meeting of Tuesday, the thirty-first day of Farvardin, one thousand three hundred and seventy-two (20 April 1993), of the Islamic Consultative Assembly and was approved by the Council of Guardians on Ordibehesht 12, 1372 (2 May 1993).

34

ANNEX 2
IRAN DECREE-LAW JULY 21, 197332
I.
The baseline, established in the Act of 22 Farvardin 1338 (12 April 1959) amending the Act of 24 Tir 1313 (15 July 1934) concerning the limits of the territorial waters and the contiguous zone of Iran, is determined as follows:

A. Straight lines joining the following points:

(1) *Point 1,* situated at the point where the thalweg of the Shatt El-Arab intersects the straight line joining the two banks of the mouth of the Shatt El-Arab at the low-water line.

(2) *Point 2,* situated at the mouth of the Behregan, whose geographical co-ordinates are: latitude 29°59'50"N., and longitude 49°33'55"E.

(3) *Point 3,* situated on the south coast of Kharg Island, whose geographical co-ordinates are: latitude 29°12'29"N., and longitude 50°18'40"E.

(4) *Point 4,* situated on the south coast of Nakhilu Island, whose geographical coordinates are: latitude 27°50'40"N., and longitude 51°27'15"E.

(5) *Point 5,* situated on Lavan Island, whose geographical co-ordinates are: latitude 26°47'25"N., and longitude 53°13'00"E.

(6) *Point 6,* situated on the south-west coast of Kish Island, whose geographical coordinates are: latitude 26°30'55"N., and longitude 53°55'10"E.

(7) *Point 7,* situated on the south-east coast of Kish Island, whose geographical coordinates are: latitude 26°30'10"N., and longitude 53°59'20"E.

(8) *Point 8,* situated at Ras-o-Shenas, whose geographical co-ordinates are: latitude 26°29'35"N., and longitude 54°47'20"E.

(9) *Point 9,* situated on the south-west coast of Qeshm Island, whose geographical coordinates are: latitude 26°32'25"N., and longitude 55°16'55"E.

32 Approved by the Council of Ministers on July 21, 1973 and transmitted by the Permanent Mission of Iran to the United Nations by a Note Verbale on December 28, 1978. Published in the United Nations Legislative

Series, ST/LEG/SER.B/19, p.55 (French text).A copy of the map referred to in section II was handed over to

the Office of Ocean Affairs, United Nations by an official of the Islamic Republic of Iran in September 1988.

35

(10) *Point 10,* situated on the south coast of Hengam Island, whose geographical coordinates are: latitude 26°36'40"N., and longitude 55°51'50"E.

(11) *Point 11,* situated on the south coast of Larak Island, whose geographical coordinates are: latitude 26°49'30"N., and longitude 56°21'50"E.

(12) *Point 12*, situated on the east coast of Larak Island, whose geographical coordinates are: latitude 26°51'15"N., and longitude 56°24'05"E.

(13) *Point 13*, situated on the east coast of Hormuz Island, whose geographical coordinates are: latitude 27°02'30"N., and longitude 56°29'40"E.

(14) *Point 14*, whose geographical co-ordinates are: latitude 27°08'30"N., and longitude 56°35'40"E.

(15) *Point 15*, whose geographical co-ordinates are: latitude 25°47'10"N., and longitude 57°19'55"E.

(16) *Point 16*, whose geographical co-ordinates are: latitude 25°38'10"N., and longitude 57°45'30"E.

(17) *Point 17*, whose geographical co-ordinates are: latitude 25°33'20"N., and longitude 58°05'20"E.

(18) *Point 18*, whose geographical co-ordinates are: latitude 25°24'05"N., and longitude 59°05'40"E.

(19) *Point 19*, whose geographical co-ordinates are: 25°23'45"N., and longitude 59°35'000"E.

(20) *Point 20*, whose geographical co-ordinates are: latitude 25°19'20"N., and longitude 60°12'10"E.

(21) *Point 21*, whose geographical co-ordinates are: latitude 25°17'25"N., and longitude 60°24'50"E.

(22) *Point 22*, whose geographical co-ordinates are: latitude 25°16'36"N., and longitude 60°27'30"E.

(23) *Point 23*, whose geographical co-ordinates are: latitude 25°16'20"N., and longitude 60°36'40"E.

(24) *Point 24*, whose geographical co-ordinates are: latitude 25°03'30"N., and longitude 61°25'00"E.

36

(25) *Point 25*, situated at the point of intersection of the meridian 61°37'03"E., and the straight line joining the shorelines at the entrance of the Gwadar Gulf at the low-water line.

B. Between points 6 and 7, situated on Kish Island, points 11 and 12, situated on Larak Island, and points 14 and 15, situated in the Strait of Hormuz, the low-water line shall constitute the baseline.

II.

The baseline used for measuring the breadth of the territorial sea of Iran is shown on the Map of the Persian Gulf the first edition of which was published in Shahrivar 1349 (September 1970) by the National Geographical Organization of Iran, on a scale of 1:1,500,000, and is attached to the present Decree. The original of the Decree is kept in the Office of the President of the Council of Ministers.

37

ANNEX 3
United States Note to the United Nations
January 11, 1994 (USUN 3509/437)

The Permanent Mission of the United States of America to the United Nations presents its compliments to the United Nations and has the honor to advise that the Government of the United States of America has studied carefully the legislative acts of Iran setting forth Iran's maritime claims, including the Act on the Marine Areas of the Islamic Republic of Iran in the Persian Gulf and Oman Sea of May 2, 1993, and Decree-Law No. 2/250-67, 31 Tir 1352 [July 22, 1973] of the Council of Ministers, taking into account the relevant provisions of international law as reflected in the 1982 United Nations Convention on the Law of the Sea, which will enter into force on November 16, 1994.

The United States is of the view that certain provisions of these acts are inconsistent with international law, and the United States reserves its rights and the rights of its nationals in that regard.

The United States wishes to recall that, as recognized in customary international law and as reflected in the 1982 United Nations Convention on the Law of the Sea, except where otherwise provided in the Convention, the normal baseline for measuring the breadth of the territorial sea is the low-water line along the coast as marked on large-scale charts officially recognized by the coastal state. Only in localities where the coastline is deeply indented and cut into, or if there is a fringe of islands along the coast in its immediate vicinity, may the coastal state elect to use the method of straight baselines joining appropriate points in drawing the baseline from which the breadth of the territorial sea is measured.

The United States notes that, notwithstanding the fact that the Iranian coastline is rarely deeply indented or fringed by islands, Iran has employed straight baselines along most of its coastline and that, in the vicinity of most segments, the Iranian coastline is quite smooth. Consequently, the appropriate baseline for virtually all of the Iranian coast in the Persian Gulf and the Gulf of Oman is the normal baseline, the low-water line. While the Convention does not set a maximum length for baseline segments, many of the segments set out in Iranian law are excessively long. In fact, eleven of the 21 segments are between 30 and 120 miles long. The United States believes that the maximum length of an appropriately drawn straight baseline segment normally should not exceed 24 nautical miles. The United States also wishes to recall that islands may not be used to define internal waters, except for situations where the islands are part of a valid straight baseline system, or of a closing line for a juridical bay. Article 3 of the 1993 Marine Areas Act of Iran asserts that the waters between islands belonging to Iran where the distance of such islands does not exceed 24 nautical miles form part of the internal waters of Iran. This claim has no basis in international law. The United States notes that Article 19[2][h] of the 1982 Law of the Sea Convention provides that "any act of wilful and serious pollution contrary to this Convention" may be considered prejudicial to the peace, good order or security of the 38 coastal state. In specifying activities in its territorial sea that Iran does not consider to be innocent, article 6 [g] of the 1993 Marine Areas Act includes "any act of pollution of the marine environment contrary to the rules and regulations of the Islamic Republic of Iran."

The United States assumes that the relevant Iranian rules and regulations will conform to the accepted rule of international law set out in Article 19 [2][h] of the 1982 Law of the Sea Convention. The United States recalls that, under articles 21 and 24 of the 1982 Law of the Sea Convention, a coastal state may adopt laws and regulations relating to innocent passage relating to the design, construction, manning or equipment of foreign ships only if they are giving effect to generally accepted international rules or standards, and may not adopt requirements that have the practical effect of denying or impairing the right of innocent passage or of discriminating in form or in fact against the ships of any state or against ships carrying cargoes to, from or on behalf of any state.

The United States notes that Iran's claim in article 7 of the right to adopt "such other regulations as are necessary for the protection of its national interest and the proper conduct of innocent passage" cannot confer upon it any greater rights than those authorized under international law.

The United States also notes that international law permits a coastal state to suspend temporarily in specified areas of its territorial sea the innocent passage of foreign ships if such suspension is essential for the protection of its security, and that such suspension may take effect only after having been duly published. Article 8 of Iran's 1993 Marine Areas Act cannot be accepted as removing the requirements that any suspension of innocent passage through parts of its territorial sea be temporary and that it take effect only after being duly published.

Article 9 of the 1993 Marine Areas Act impermissibly seeks to require foreign warships, and vessels carrying dangerous or noxious substances harmful to the environment, to obtain prior authorization from Iran to pass through Iran's territorial sea. Such a requirement has no foundation in the provisions of the 1982 Law of the Sea Convention, and the United States will continue to reject, as contrary to international law, any attempt to impose such a requirement on the exercise of the right of innocent passage of all ships.

The United States assumes that Iran will not seek to exercise criminal jurisdiction, pursuant to Article 10 of the 1993 Marine Areas Act, on board ships other than merchant ships and government ships operated for commercial purposes, or to exercise civil jurisdiction, pursuant to Article 11 of this Act, in situations not contemplated by Article 28 of the 1982 Law of the Sea Convention. The United States further recalls that the scope of a coastal state's authority in its contiguous zone, a maritime zone contiguous to and seaward of the territorial sea in which freedoms of navigation and over flight may be exercised, is limited to the exercise of the control necessary to prevent and punish infringement of its customs, fiscal, immigration and 39

sanitary laws and regulations committed within its territory or territorial sea, and that the authority of the coastal state to enforce its environmental laws seaward of its territorial sea is as prescribed in Article 220 of the Convention. The claim in article 13 of the 1993 Act to adopt measures in Iran's contiguous zone necessary to prevent infringement of its security, maritime and environmental laws exceeds that permitted by international law.

Although a coastal state may establish, in accordance with Article 60, paragraph [4] and [5], of the 1982 Law of the Sea Convention, safety zones of a radius not exceeding 500 meters around artificial islands and other installations and structures located within its exclusive economic zone. International law does not authorize a coastal state to establish so-called security zones in such areas. Article 14 [b][1] or the 1993 Marine Areas Act impermissible asserts the right to do so. That provision also appears to claim more authority to control the laying of submarine cables and pipelines on Iran's continental shelf than is permitted by international law as reflected in Article 79 of the 1982 Law of the Sea Convention. Further, international law permits a coastal state to regulate only marine, scientific research in its exclusive economic zone, not "any kind of research" as claimed in article 14 [b] [2] of the 1993 Marine Areas Act. In particular, hydrographic surveys conducted seaward of the territorial sea are not marine scientific research and are not subject to coastal state jurisdiction.

The United States notes that, to the extent Article 16 of the 1993 Marine Areas Act seeks to prohibit in the Iranian exclusive economic zone the exercise by foreign warships and military aircraft of their freedoms of navigation and over flight, it contravenes international law. The United States has previously protested Iran's claim in this regard, and will continue to operate its ships and aircraft consistent with its rights under international law.

The Government of the United States wishes to assure the Government of the Islamic Republic of Iran that its objections to these claims should not be viewed as singling out the Islamic Republic of Iran for criticism, but is part of its worldwide effort to preserve the internationally recognized rights and freedoms of the international community in navigation and over flight and other related high seas uses, and thereby maintain the balance of interests reflected in the Convention.

This is only of a number of U.S. protests of those claims by coastal states which are not consistent with international law as reflected in the 1982 United Nations Convention on the Law of the Sea.

The Government of the United States requests that this Note be circulated by the United Nations as part of the next Law of the Sea Bulletin.

The Permanent Mission of the United States of America to the United Nations avails itselfof this opportunity to renew to the United Nations the assurances of its highest regard

Bahman Aghai Diba

-B.A. Political Sciences (covering legal issues), Faculty of Law and Political Sciences, **Tehran University**.

-M.A. International Relations, Center for Graduate **International Studies**, Tehran University.

-Ph.D. International Law (the Law of the Seas), Faculty of Law, **Delhi University**.

<u>**Occupations and experiences** </u>:
- **World Resources Company**, Consultant in International Affairs
- **Dr. Shirin O. Entezari & Assoc**. (International Law Firm acting in Iran, Turkmenistan, Azerbaijan), Managing partner of the Tehran Office
- **Legal Department**, Iranian Ministry of Foreign Affairs, International law expert, Contacts negotiator, Law of the Sea Expert
- Second Secretary in charge of International and Economic affairs, Special Liaison Officer with the Office of the Non-aligned Movement's Affairs, Iranian Embassy (accredited for Nepal, Burma, and Vietnam), New Delhi.

- Liaison Officer, **Asian-African Legal Consultative Organ-ization** (**AALCO**), New Delhi, India
- Member of the Iranian Delegations to the **United Nations General Assembly** sessions, including the sixth committee (Legal Affairs)
- Member of the Iranian Delegation to **IRAN-IRAQ Peace Talks**, under the supervision of the UN Secretary General, **New York** and **Geneva** (1988)
- Representative of Iran in the International Conference for Revision of IMO Convention on the Civil Liability for Oil Pollution Damage (CLC) and the Convention of International Compensation Fund, Jakarta (**Indonesia**) 1984
- Member of the Iranian Delegations to <u>AALCO</u> Annual Sessions in **New Delhi (India), Katmandu (Nepal), and Nairobi (Kenya)**
- Member of the Iranian Delegation to London for Nego-tiations on Damages to Embassy premises in London and Tehran 1987
- Working with the Office of **United Nations High Commissioner for Refugees**.
- Working with the **United Nations Information Office** in Tehran.

Publications in English:

-Law and Politics of the Caspian Sea 2006

- FAQ about the Nuclear Case of Iran, 2007

- Problems of Iran: How not to govern, 2006

Translation from English into Persian:

-**Black's Law Dictionary**, Published by Ghangedanesh Publications, Tehran, 1999

- **The International Law of the Sea,** by R R Churchill and A V Lowe, Manchester University Press, Published five times in Persian, and the selected textbook for the Law of the Sea in post graduate courses on the law of the sea. The last edition published in 2005.

- **A Modern Introduction to the International Law**, By Professor Akhurst, Published by the Iranian Bureau for International Legal Services, a university textbook for the undergraduate students of the international law, Tehran, 1997.

- **Dictionary of International Law**, By Robert Bledsoe and Boczek, Published by Ghangedanesh Publications, Tehran, 1998.

- **International law governing Communications and Information**, compiled by Edward W. Ploman, International Institute of Communications, Ghangedanesh Publications, Tehran, 200

- **The Foreign Policy of the USSR**, Tehran, 1986

-**Human rights**, published by the Office of the United Nations in Tehran, 2000

Other works:

-**The Law and Politics of the Caspian Sea in the 21st Century**, [in English] The positions and views of the Coastal states, IBEX Publishers, Bethesda, Maryland, 2003.

- **The Rise and Fall of the Pahlavi Dynasty** [in English], translation of the Memoirs of General Hussein Fardoust, Tehran, 1995, Institute for political Studies and Researches

- **Collection of 12 Articles Concerning Law of The Sea** and Iranian Problems, including the legal regime of the Caspian Sea, Iran's Position towards innocent passage in the International Straits and Territorial Waters, and the Passage of Military Ships in the Territorial Sea and The Right of Landlocked States, Published by Ghangedanesh Publications, Tehran, 1997.

- **A Terminology of Human Rights**, published by Ghangedanesh Publications, Tehran 1997.

-**Ikhvan Al Muslemin [Muslim Brotherhood] in the Middle East,** Tehran, 1986.

- Many Articles Regarding International Law and relations published in Iranian and international specialized periodicals

8427992R1

Made in the USA
Charleston, SC
08 June 2011